T0300962

Renewable and Waste-Heat Utilisation Technologies

Understand the science and engineering behind conventional and renewable heat loss recovery techniques with this thorough reference guide. This book provides you with the knowledge and tools necessary to assess the potential waste-heat recovery opportunities that exist within various industries and select the most suitable technology. In particular, technologies that convert waste heat into electricity, cooling or high-temperature heating are discussed in detail, alongside more conventional technologies that directly or indirectly recirculate heat back into the production process. Essential reading for professionals in chemical, manufacturing, mechanical and processing engineering who have an interest in energy conservation and waste-heat recovery.

Nareshkumar B. Handagama is a chemical engineer with more than 35 years of research and development and industrial experience in some of the world's largest public and private utilities, chemical and petrochemical companies. Currently, he is the chief operating officer at Sri Lanka Nano and Advanced Technology Centre (SLINTEC). He is a licensed professional engineer (PE) in the USA, and a chartered engineer in the United Kingdom, a Fellow of the American Institute of Chemical Engineers (FAIChE) and a fellow of the Institution of Chemical Engineering (FIChemE, London, UK).

Martin T. White is a senior lecturer in Mechanical Engineering and member of the Thermo-Fluid Mechanics Research Centre at the University of Sussex.

Paul Sapin is a post-doctoral research associate and a leader of the Energy Division in the Clean Energy Processes (CEP) Laboratory at Imperial College London.

Christos N. Markides is a professor of Clean Energy Technologies at Imperial College London where he leads the Clean Energy Processes (CEP) Laboratory and coordinates the Experimental Multiphase Flow (EMF) Laboratory. He is also a co-founder and director of the recent spin-out company Solar Flow.

Renewable and Waste-Heat Utilisation Technologies

Thermal Energy Recovery, Conversion, Upgrading and Storage

NARESHKUMAR B. HANDAGAMA

Sri Lanka Institute of Nanotechnology

MARTIN T. WHITE

University of Sussex

PAUL SAPIN

Imperial College London

CHRISTOS N. MARKIDES

Imperial College London

CAMBRIDGE
UNIVERSITY PRESS

CAMBRIDGE
UNIVERSITY PRESS

Shaftesbury Road, Cambridge CB2 8EA, United Kingdom

One Liberty Plaza, 20th Floor, New York, NY 10006, USA

477 Williamstown Road, Port Melbourne, VIC 3207, Australia

314–321, 3rd Floor, Plot 3, Splendor Forum, Jasola District Centre, New Delhi – 110025, India

103 Penang Road, #05–06/07, Visioncrest Commercial, Singapore 238467

Cambridge University Press is part of Cambridge University Press & Assessment,
a department of the University of Cambridge.

We share the University's mission to contribute to society through the pursuit of
education, learning and research at the highest international levels of excellence.

www.cambridge.org
Information on this title: www.cambridge.org/9781108480772
DOI: 10.1017/9781108691093

© Nareshkumar B. Handagama, Martin T. White, Paul Sapin and Christos N. Markides 2023

First published 2023

A catalogue record for this publication is available from the British Library.

ISBN 978-1-108-48077-2 Hardback

Contents

Acronyms

Roman Symbols

A	heat-transfer area, m^2
c	velocity, m/s
c_p	specific-heat capacity, J/(kg K)
C	cost, \$
C_0	total investment cost, \$
C_c	cost of electricity, \$/kWh
C_g	cost of natural gas, \$/kWh
$C_{o\&m}$	operation and maintenance costs, \$/kWh
d_h	hydraulic diameter, m
D	diameter, m
D_s	specific diameter
$\mathcal{D}$	thermal diffusivity, m^2/s
e_{th}	thermal effusivity, $J/m^2/s^{1/2}/K$
f	friction factor
F	Martinelli parameter
g	acceleration due to gravity, m/s^2
h	enthalpy, J/kg
k	thermal conductivity, W/(m K)
L	length, m
LCOE	levelised cost of electricity, \$/kWh
m	mass, kg
$\dot{m}$	mass-flow rate, kg/s
n	operating hours per annum
N_s	specific speed
Nu	Nusselt number
NPV	net-present value, \$
P	pressure, Pa
ΔP	pressure drop, Pa
P_r	reduced pressure
PB	payback period, years
PP	heat-exchanger pinch point, K
Pr	Prandtl number

q	heat transfer per unit mass, J/kg, vapour quality
Q	heat transfer, J
$\dot{Q}$	heat-transfer rate, J/s, volumetric flow rate m^3/s
r	discount rate, %
Re	Reynolds number
s	entropy, J/(kg K)
S	annual savings, $
SIC	specific investment cost, $/kW
t	time/year
T	temperature, K
ΔT_k	endo-reversible heat-pump temperature difference, K
ΔT_{log}	log-mean temperature difference, K
ΔT_{sh}	amount of superheat K
w	specific work, J/kg
W	work, J
$\dot{W}$	power, W
U	internal energy, J, overall heat-transfer coefficient W/(m^2 K)
x	fluid composition
X	exergy, J
$\dot{X}$	exergy rate, J/s
z	height, m

Greek Symbols

α	heat-capacity ratio, heat-transfer coefficient, W/(m^2 K)
β	heat-conductance ratio
ε	heat-exchanger effectiveness
η	thermal efficiency/isentropic efficiency
θ	non-dimensional heat source temperature drop
μ	dynamic viscosity, Pa s
ρ	density, kg/m^3
ϕ	coefficient of performance (power-driven)
ψ	coefficient of performance (heat-driven)
ω	rotational speed, rad/s

Subscripts

0	dead state
c	cold
ch	chiller
ci	cold-fluid inlet
cp	cold-fluid pinch
co	cold-fluid inlet
cr	critical point
e	expander
ev	evaporator
h	hot

hi	hot-fluid inlet
hp	hot-fluid pinch, heat pump
ho	hot-fluid outlet
i	inner
l	saturated liquid
min	minimum
max	maximum
n	net
o	outer
p	pump
ph	preheat
s	conditions after isentropic expansion
sh	superheater
v	saturated vapour
wf	working fluid
′	saturation conditions

1   Introduction

Growing concerns over climate change, environmental pollution and energy security have demanded society to re-evaluate the ways in which energy is generated and consumed. This encompasses the generation of clean renewable energy, in addition to reducing the total energy consumption. Arguably, much of the current renewable revolution is linked to renewable technologies, such as solar photovoltaic and wind turbines, that generate electricity directly from the environment. However, to focus only on these technologies is to neglect the fact that the provision of heat is equally, if not more, important than electricity generation. Heat is one of the most basic requirements for life, and is needed for cooking, washing and keeping warm. Alongside this, heat is required in a wide range of industrial processes across a range of temperatures that could range anywhere between a few tens of degrees Celsius up to a few thousand. Considering this, it is unsurprising that roughly a third of global primary energy supply is used for heat production, whilst the production of heat accounts for over half of global energy consumption. This heat could come from a variety of sources, including primary fossil fuels such as natural gas, coal or oil, or from renewable sources, including biomass, geothermal or solar energy. Thus, irrespective of the source of the heat, technologies that can utilise this heat, either for direct heating purposes, upgrading it to a higher temperature, or for its conversion into electricity or cooling, have an important role in a future clean and sustainable society.

Alongside reducing consumption and generating renewable energy, there is also the need to strive for improved energy efficiency, which means developing new technologies or improving existing ones, so that much of the total energy that is consumed is converted to useful energy that can be used for either heating, cooling or electricity generation. Thus, a growing and important area of interest is waste-heat recovery. Currently, many energy-intensive industries reject a relatively large proportion of the energy they consume to the environment in the form of waste heat. For example, Johnson and Choate (2008) state that, as of 2008, the industrial sector in the United States consumed approximately one-third of all the primary energy used annually, corresponding to 9,400 TWh. This corresponds to the emission of 1,680 million metric tons of carbon dioxide equivalent, along with emissions harmful to both human health and the environment. However, as much as 20-50% of the energy consumed is ultimately lost to the environment in the form of waste heat. Many existing efforts to improve the energy efficiency of industrial processes have focussed on reducing the energy consumption of the equipment used during the manufacturing process by improved component design and plant operation, or through the modification of the manufacturing process. In contrast, the implementation of technologies to capture and reuse waste heat has not received as much attention. Historically, this is because there has been little incentive for industry to recover waste heat due to there being either limited use for it on-site or high-associated costs for waste-heat recovery technologies. Furthermore, Johnson and Choate (2008) also observed that in the United States, 90% of waste heat is below 316 °C, whilst 60% is below 230 °C. These temperatures fall somewhat short of the high-temperatures required for conventional heat to power applications, or high-temperature process heating.

However, over recent years there has been a significant interest in introducing technologies to make many industrial processes more energy efficient. In particular, the amount of waste heat from industries such as glass; iron and steel; cement; oil and gas; and food and drink have become particular areas of focus (Johnson & Choate 2008; McKenna 2009). Within an industrial process, waste heat could be recovered at multiple points within the plant. This waste heat could be in the form of hot exhaust gases directly leaving an industrial process; hot cooling-water streams; or through conduction, convection and radiation from the hot surfaces of process equipment or the final manufactured product. This waste heat could be recirculated back into the production process, either through direct or indirect heat exchange, or converted into other forms of energy using a suitable technology. It is these heat conversion technologies that are the primary focus of this textbook.

1.1   Heat Engines

In general, heat engines, or heat to power technologies, convert heat into mechanical shaft power, which in turn can be converted into electricity using a generator. Heat engines are widely used in power generation, in which the heat is typically provided from the combustion of fossil fuels; common examples are those used in coal- and gas-fired power plants. However, in the context of this textbook, heat could be supplied from any renewable heat source, or be in the form of waste heat from an existing engine or waste heat that is discarded from an existing industrial process. The efficiency of the conversion process is very much dependent on the temperature of the available heat, with higher temperatures corresponding to higher conversion efficiencies – a higher conversion efficiency meaning we are able to convert a larger percentage of the energy contained in the heat source into mechanical power. Therefore, in many applications where the available heat is at a sufficiently high temperature, well-established heat engine technologies can be used to generate electricity. However, for lower temperature heat sources, typically below a few hundred degrees Celsius, the performance of conventional heat engines diminishes. Thus, alternative technologies for the conversion of this lower-grade heat into power are required. This is currently an active area of research and development, and one can expect that as these technologies mature, it will become more and more economical to convert lower-grade waste heat into electricity. Thus, low-temperature heat engines will be one of the key technologies discussed in this book.

1.2   Heat Pumps and Chillers

Heat pumps and chillers are a family of technologies that are also of when considering heat utilisation. In general, there are two groups of systems. The first is the vapour-compression heat pump, or refrigeration cycle, which can be used to transfer heat from low-temperature heat source to a higher-temperature heat sink, whilst the second are absorption or adsorption systems that can be used to provide cooling using an available heat source. Both technologies are of interest in this book.

A vapour-compression system can be used to transfer heat from a low-temperature heat source to a higher-temperature heat sink, and in general can be thought of as the opposite of a heat engine in that this increase in the value (or quality) of the heat requires work to be added to the system (in comparison, a heat engine generates work and, in the process, transfers heat from a high-temperature heat source to a low-temperature heat sink). Currently, heat pumps are widely used in applications such as ground- and air-source heat pumps, alongside refrigerators and air conditioning systems, although in the latter two applications they are not typically referred to as a heat pump but instead a refrigeration cycle. This distinction comes about depending on whether value is attached to the low-temperature heat source or the high-temperature heat sink. For example, consider an air-source heat pump and a common domestic refrigerator, which are common examples of this technology. The former extracts heat from the ambient air, upgrades this to a higher temperature through compression and then uses this heat for space or water heating, whilst the latter extracts heat from inside the refrigerator, upgrades this heat to a higher temperature through compression and then rejects this heat to the ambient surroundings. Thus, both operate on exactly the same principle, but the former is referred to as a heat pump, whilst the latter is referred to as a refrigeration cycle. This is because in the case of the air-source heat pump we are interested in the heat generated from the system, whilst in the case of the refrigerator we are interested in the amount of heat that can be extracted from the heat sink. Absorption- and adsorption-based systems, on the other hand, use heat to drive a thermodynamic cycle that is able to generate a cooling effect without requiring significant compression. They are most suitable for applications where there is a renewable heat source available and a demand for cooling.

Heat pump technology is widely used within refrigeration systems, and commercial products are available to convert renewable heat extracted from the ground, the ambient air or a water source into space and water heating. Similarly, absorption and adsorption systems are available to convert renewable heat, such as solar energy, into cooling. However, both technologies can also find use in waste-heat recovery applications. In this capacity, heat pumps are a suitable technology to upgrade waste heat to a higher temperature, and then inject this heat into the production process, whilst absorption and adsorption systems may be adopted to convert available waste heat into cooling, where there is an on-site demand. Thus, both technologies are of relevance with regards to the utilisation of both renewable and waste heat and will be the second group of technologies that will be discussed in this book.

1.3   Classification of Heat

To classify heat, it is first necessary to understand what heat is. Any object, or system, at a temperature above absolute zero will contain an amount of thermal energy, which is energy that could be extracted if that object or system was cooled down to absolute zero. Therefore, there is a clear link between the thermal energy contained within a system and its temperature. Thus, thermal energy can be thought of as the energy contained within a system that is responsible for its temperature. In comparison, heat transfer is the flow of thermal energy from one reservoir to another. For example, let's consider the scenario where we have two reservoirs, and one reservoir is held at a higher temperature than the other. Assuming that the two reservoirs are able to interact, there will be a flow of thermal energy from the hot reservoir to the colder reservoir. It is this flow of energy that is referred to as heat, and heat will always flow from a hot reservoir to a cold reservoir. Heat can be transferred via three modes: conduction, convection and radiation. Conduction is the transfer of heat through an object, which may be either a solid, liquid or gas, at the molecular level, and occurs when that object is exposed to a temperature gradient. Convection is the transfer of heat through either a liquid or a gas and occurs due to the motion of the fluid. This can be random molecular motion, which is referred to as diffusion, or the bulk motion of the fluid. Finally, radiation is the transfer of heat via electromagnetic waves, and occurs between two objects even when there is no medium in between them to carry the heat via conduction or convection.

From what has been discussed, it follows that larger the temperature difference between a hot and a cold reservoir, more the heat available. Thus, the amount of heat available depends on both the size of a reservoir and its temperature. Therefore, depending on the reservoir, heat can be provided at a range of different temperatures. It also follows that heat is required at a range of different temperatures. This is in contrast to forms of energy, such as electricity, where no such distinction between different types of electricity is made. Alongside this distinction, it is also worth noting that, unlike electricity, heat cannot be easily transported across large distances. Both of these aspects will have impact on how any available heat is used, and the technologies that may be used to convert that heat into other forms of energy. Considering that heat may be available at different temperatures, one way to classify heat is to categorise it according to its temperature. In this book, heat is commonly referred to as either low-temperature, medium-temperature or high-temperature heat. However, the temperatures to which each classification is related are not clearly defined. Within this textbook, we will adhere those ranges, as summarised in Table 1.1.

Heat can also be classified according to its origin. For example, heat may either be from the combustion of fossil fuels, from a renewable source or from waste heat. A summary of potential heat sources, alongside their main advantages, disadvantages and temperature ranges, is provided in Table 1.2.

From this it easy to see why fossil fuels are so widely used given their ability to generate heat across a wide range of temperatures and given the fact they are compact, stable and easy to transport. Alternatively, biomass, which is a combustible

Table 1.1 Classification of heat according to its temperature

Classification	Temperature range
Low-temperature heat	< 100 °C
Medium-temperature heat	100–400 °C
High-temperature heat	> 400 °C

Table 1.2 Summary of possible heat sources

Source	Maximum Temperature	Advantages	Disadvantages
Fossil fuels (coal, oil, gas)	> 1,000 °C	– Readily available – Cheap – Large temperature range – High caloric value – Stable – Easy to transport	– Global warming – Smoke/air pollution – Finite supply
Biomass	> 1,000 °C	– Good fossil fuel alternative – Large temperature range – Stable – East to transport	– Deforestation – Smoke/air pollution – Competition with food production
Geothermal	< 300 °C	– Renewable and free – Independent of season – Independent of weather	– Geographically limited – High upfront costs – Low-temperature heat
Solar	> 1,000 °C	– Largest renewable resource – Renewable and free – No pollution/global warming – Simple construction	– Large seasonal variance – Large daily variance – Provides heat only when there is sun
Waste heat	Application dependent	– Free – Improve plant efficiency	– Process dependent – Variable/intermittent – Recovery process may interfere with plant

organic matter derived from plants, is a promising replacement of fossil fuels, but is still associated with environmental pollution and could lead to further deforestation if the organic matter is not provided on a renewable basis. Geothermal energy, in which heat is extracted from deep under the ground, is also a promising candidate for renewable heat production, providing a steady supply of heat that is not dependent on weather or season. However, temperatures may be limited, whilst its implementation is limited to areas where there is accessible geothermal supply. Solar energy is one of the most promising technologies for the provision of heat. Depending on whether relatively low-cost solar collectors, such as flat plate or evacuated tubes, or higher-cost collectors, such as parabolic troughs, solar dishes and solar towers, are employed, heat can be generated across a range of temperatures covering applications including hot water and space heating, industrial process heating and concentrated solar power plants. However, the main disadvantage is the large daily and seasonal variances in availability.

Alongside these heat sources, heat may be available in the form of waste heat from an industrial process, or as waste heat from an existing heat engine. In this case, this waste heat represents a free heat source that can be utilised to improve plant efficiency. The specific nature of the heat available will depend on many factors, including whether the heat is available in the form a solid, liquid or gas; whether the upstream process is continuous, cyclic or intermittent; its load factor; the accessibility of the heat source; and whether the industrial process puts any constraints on the waste-heat recovery process.

1.4   Importance of Low- and Medium-Temperature Heat

As noted earlier in the chapter, heat is one of the most basic human requirements, and each one of us uses heat daily to wash, cook and stay warm. For these purposes, heat is used directly, and is generally required at low temperatures. However, for the conversion of heat, derived from the combustion of fossil fuels, into electricity, it is advantageous to generate heat at higher temperatures, which enables high conversion efficiencies to be achieved. Thus, historically, developments in heat engine technology have focussed on increasing the maximum temperature of operation as much as possible. Alongside this, where economically feasible, efforts have been made to further increase efficiency by using any heat that is rejected from the heat engine for additional purposes. One example would be a fossil fuel-driven combined cycle, where the heat derived from the combustion process drives a gas turbine, and the heat rejected from the gas turbine is then used to drive a steam cycle. Another would be a combined heat and power system, where the heat rejected from the heat engine is used for the provision of hot water.

However, as we move from fossil fuels to more sustainable heat sources, and further improve energy efficiency, it will become more and more critical to employ technologies that are capable of utilising low- and medium-temperature heat. To extend the analogy of the combined cycle, from a technical point of view an additional heat engine could be installed to convert the heat rejected from the steam cycle into additional electricity. To date, they may be of limited economic benefit in doing this, but as more pressure is placed on energy producers and industry to reduce their environmental impact, more onus will be placed on recovering this lower-grade heat.

The importance of recovering and utilising low- and medium-grade heat is already well documented within the literature. For example, we have already seen that over 90% of waste heat in the United States is below 316 °C (Johnson & Choate 2008), whilst in China over 54% of waste heat that is available is at temperatures below 500 °C (Ma et al. 2012). Another report estimates that of the waste heat that is technically feasible to recover from manufacturing industries within the United States, 37% is below 230 °C (Elson, Hampson & Tidball 2015). Similarly, within Europe, Bianchi et al. (2019) estimated that there is a total waste-heat recovery potential of

928 TWh, of which 51% is below 100 °C, 19% is between 100 and 300 °C and the remaining 30% above 300 °C. An earlier study by Campana et al. (2013a) estimated that the utilisation of low- and medium-temperature for power generation within European energy-intensive industries could save approximately 2 billion euros, and over 8 million tonnes of harmful gas emissions, per annum. Clearly then, the recovery and utilisation of low- and medium-temperature waste heat is of great importance.

Alongside this, it is also worth considering the renewable heat sources listed in Table 1.2. Compared to fossil fuels, geothermal sources are available at much lower temperatures. Furthermore, whilst solar energy can be concentrated to achieve temperatures as high as 1,000 °C, the solar collectors required to achieve this are generally high cost and employed for large-scale systems. However, for smaller-scale systems, which may be employed within a distributed power generation system, or even for a single dwelling, it is likely that these systems would rely on lower-cost solar collectors that provide heat at a much lower temperature. Thus, as we move towards a higher utilisation of renewable heat sources, technologies that are capable of utilising this low- and medium-temperature heat will become more and more important.[1]

1.5 Objective and Structure of This Book

From this brief introduction, it should be clear to the reader that technologies to convert renewable and waste heat into other forms of energy have an important role to play in a clean and sustainable society. Clearly, if a renewable or waste-heat stream is available, and there is a demand for energy in various forms, there are energy savings to be made by installing a suitable heat-utilisation technology. Furthermore, there may also be significant economic and environmental benefits to installing such technologies, provided that these technologies are implemented correctly. Therefore, the identification of the optimal pathways for the utilisation of renewable and waste-heat streams is of paramount importance, and the identification of this optimal pathway is centred around three essential components, namely: (i) identification of an accessible source of heat; (ii) identification of a suitable heat utilisation technology; and (iii) identification of a use for the recovery energy.

With this in mind, the objective of this book is to provide the reader with the knowledge and tools necessary to assess the potential heat-utilisation opportunities that exist, with an emphasis on waste-heat recovery. This includes demonstrating how to examine the quantity and quality of the heat available, an overview of the available heat-utilisation technologies and the formulation of modelling tools that can be used to design, optimise and select a heat-utilisation technology for a particular heat stream. It is hoped that these discussion points will aid in the identification of future strategies that can lead to a more widespread implementation of heat-utilisation

---

[1] Of course, utilising lower-temperature heat will result in lower conversion efficiencies compared to fossil fuel-based systems. However, since renewable heat sources are essentially free, there is no direct economical penalty to requiring more heat to generate the same amount of electricity.

technologies, and ultimately lead to increased energy efficiency within a number of important sectors, and reduce their impact on the environment.

This book is divided into nine chapters. In Chapter 2, a review of the fundamental scientific theory required to evaluate heat-utilisation technologies is provided. This chapter should provide the reader with the basic understanding that is necessary to conduct preliminary estimates of the potential performance of different heat-utilisation technologies based on the heat source available. In Chapter 3, a summary of heat-utilisation technologies is provided, covering the three main areas of heat engines, heat pumps and absorption/adsorption chillers. Alongside describing the operating principle of each technology and highlighting research trends, an overview of commercially available systems is also provided. In Chapter 4, a set of technology-agnostic models are presented for the three different systems considered within this book. These models are helpful in providing a first-stage assessment of the feasibility of heat engines, heat pumps and chillers installation for a particular application, without requiring a detailed knowledge of the internal working of the system. This is followed by a much more detailed treatment of these technologies in Chapter 5 and 6, respectively. Specifically, Chapter 5 and 6 discuss important aspects related to the detailed design, simulation and optimisation of these systems, including both performance and cost considerations. In Chapter 7, a review of potential applications for heat-utilisation technologies is included. In particular, this covers waste-heat recovery and utilisation opportunities within a range of industrial sectors, in addition to the opportunities to convert renewable heat into other forms of energy. Finally, in Chapter 8, the role of thermal energy storage systems is discussed, which are important when considering that a large number of waste-heat or renewable-heat streams may be variable or intermittent in nature.

2   Heat-Recovery Fundamentals

The aim of this chapter is to equip the reader with the necessary tools to evaluate the potential for waste-heat recovery at a particular site and to estimate the expected thermodynamic performance of different waste-heat recovery technologies. The content covered in the chapter includes: (i) an overview of fundamental thermodynamic principles and their applications to waste-heat recovery; (ii) methods to conduct a first-stage assessment of heat engine, heat pump and absorption-chiller thermodynamic performance; (iii) heat-exchanger fundamentals; and (iv) a discussion of important considerations when evaluating a potential waste-heat stream.

2.1   Fundamental Thermodynamics

2.1.1   First Law of Thermodynamics

The first law of thermodynamics is a form of the conservation of energy law, and arises from the fact that energy cannot be created or destroyed, but instead it changes from one form to another. Mathematically it is expressed as follows:

$$Q - W = \Delta U. \tag{2.1}$$

It says that the amount of heat added to a system Q minus the work done by the system W is equal to the change in the internal energy of the system ΔU. The internal energy of a system is comprised of thermal energy, or enthalpy, kinetic energy and gravitational potential energy, and is given as:

$$U = m \left(h + \frac{c^2}{2} + gz \right), \tag{2.2}$$

where m is the mass of the system, h is the specific enthalpy, c is the velocity, g is the acceleration due to gravity and z is the height. Therefore, considering the scenario in which a system undergoes a change from state 1 to state 2, one can derive the following:

$$Q - W = \Delta U = U_2 - U_1 = m \left[(h_2 - h_1) + \frac{(c_2^2 - c_1^2)}{2} + g(z_2 - z_1) \right]. \tag{2.3}$$

Typically, when we analyse a waste-heat recovery technology, the change in height is negligible, whilst the velocity of the fluid is small. Furthermore, the heat addition, work output and mass are given as a rate (i.e. per unit time). Therefore, Equation 2.3 reduces to:

$$\dot{Q} - \dot{W} = \dot{m}(h_2 - h_1), \tag{2.4}$$

where $\dot{Q}$ is the rate of heat addition with units of J/s, $\dot{W}$ is the power output with units of J/s and $\dot{m}$ is the mass flow rate with units of kg/s.

2.1.2　Heat Potential

Using the first law of thermodynamics it is now possible to determine the maximum amount of heat available within a given waste-heat stream. Let's assume we have a waste-heat stream leaving an industrial process at a known pressure P_1, and temperature T_1, with a known mass flow rate $\dot{m}$. The specific enthalpy of the waste-heat stream is first determined using an appropriate equation of state:

$$h_1 = f(P_1, T_1). \tag{2.5}$$

The maximum amount of heat can be extracted when this waste-heat stream is cooled down as much as possible to the minimum allowable temperature $T_{2,min}$, which, in most cases, will occur when the waste-heat stream is cooled down to the ambient temperature. Assuming no pressure drop within the system (i.e. $P_2 = P_1$), the minimum specific enthalpy of the waste-heat stream after the heat-recovery process will be:

$$h_{2,min} = f(P_2 = P_1, T_{min}). \tag{2.6}$$

Therefore, the maximum heat potential $\dot{Q}_{max}$ is given as:

$$\dot{Q}_{max} = \dot{m}(h_{2,min} - h_1), \tag{2.7}$$

or, since $h_{2,min} < h_1$:

$$-\dot{Q}_{max} = \dot{m}(h_1 - h_{2,min}), \tag{2.8}$$

where the negative sign indicates that heat is leaving the system, rather than entering the system.

2.1.3　Second Law of Thermodynamics

The second law of thermodynamics accounts for the fact that a real thermodynamic process is not reversible. For example, it is intuitive that heat will always flow from a hot reservoir to a cold reservoir, and never in reverse, unless external work is applied to the system. Irreversibility is measured by the quantity entropy S, and expressed mathematically, the second law states that the change in entropy of a system ΔS is given as:

$$\Delta S = \int \frac{dQ}{T}, \tag{2.9}$$

where dQ is an incremental transfer of heat into the system and T is the absolute temperature of the system. For a process to be reversible, $\Delta S = 0$, whilst in a real thermodynamic process $\Delta S > 0$, meaning that entropy will also increase.

2.1.4 Heat Engines

A heat engine is a technology that converts heat into power by absorbing heat from a high-temperature heat source, reducing this heat to a lower temperature state and then rejecting heat to a cold-temperature heat sink. A heat engine typically operates within a closed loop, and therefore the internal energy of the system at the start and end of the loop is equal. Applying the first law of thermodynamics to the system, we have:

$$Q_h - Q_c = W, \tag{2.10}$$

where Q_h is the heat absorbed by the system from the high-temperature heat source, Q_c is the heat rejected to the low-temperature heat sink and W is the work produced by the heat engine.

The thermal efficiency of a heat engine η is defined as the ratio of the amount of work generated by the engine to the amount of heat added to the system, therefore:

$$\eta = \frac{W}{Q_h} = \frac{Q_h - Q_c}{Q_h} = 1 - \frac{Q_c}{Q_h}. \tag{2.11}$$

If we assume the heat engine to be ideal, and therefore reversible, from the second law of thermodynamics we have:

$$\frac{Q_h}{T_h} = \frac{Q_c}{T_c}, \tag{2.12}$$

where T_h and T_c are the heat-source and heat-sink temperatures, respectively. From this it follows that:

$$\eta = 1 - \frac{T_c}{T_h}, \tag{2.13}$$

which is referred to as the Carnot efficiency of a heat engine, and represents the theoretical maximum efficiency that can be obtained by a heat engine operating between a known heat-source and heat-sink temperature.

For a more realistic estimate of heat engine performance, accounting for irreversibilities within the heat engine, Novikov (1958) derived a simple relationship for the optimal thermal efficiency of a heat engine, and this is referred to as the endo-reversible efficiency:

$$\eta = 1 - \sqrt{\frac{T_c}{T_h}}. \tag{2.14}$$

In deriving this expression, Novikov expressed the power from a heat engine as the product of the heat transferred from the heat source to the working fluid of the thermodynamic cycle, which is at a mean temperature T_p and below T_h (i.e. $Q_h \propto T_h - T_p$), and the thermal efficiency of the thermodynamic cycle (i.e. $\eta = 1 - T_c/T_p$). Then,

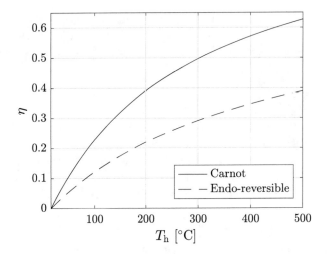

Figure 2.1 The effect of the heat-source temperature on the Carnot and endo-reversible efficiencies of heat engine. The heat-sink temperature is fixed at 15 °C

by differentiating this expression with respect to the mean working fluid temperature, he was able to determine the optimal value for T_p that corresponds to the maximum power output from the cycle.

Both the Carnot and endo-reversible efficiencies have been plotted as a function of heat-source temperature in Figure 2.1, assuming a fixed heat-sink temperature of 15 °C. This figure is useful for two reasons. Firstly, the results can be used to provide a first-stage assessment of the expected thermal efficiency of a heat engine based on a known waste-heat-source temperature. Secondly, the results demonstrate the significant reduction in thermal efficiency as the heat-source temperature reduces, highlighting a significant challenge when considering the utilisation of low-temperature waste-heat sources.

Example 2.1   A waste-heat stream of pressurised water at 150 °C, with a mass-flow rate of 1.5 kg/s and specific-heat capacity of 4.2 kJ/(kg K), is cooled down to 100 °C, and this heat is converted into electricity using a heat engine. The heat engine rejects heat to a heat sink which is maintained at a constant temperature of 15 °C. Estimate the power produced from a Carnot heat engine, and from a more realistic endo-reversible heat engine.

Solution

Firstly, calculate the amount of heat that is transferred into the heat engine:

$$\dot{Q}_h = \dot{m}_h c_{p,h}(T_{hi} - T_{ho})$$
$$\dot{Q}_h = 1.5 \times 4.2 \times (150 - 100) = 315 \text{ kW}.$$

The thermal efficiencies of a Carnot engine and an endo-reversible heat engine can be calculated from the heat-source and heat-sink temperatures. However, the heat-source undergoes a temperature reduction and is not constant. In this instance, a reasonable approach is to use the average heat-source temperature.

$$T_{h,av} = 0.5(T_{hi} + T_{ho}) = 0.5(150 + 100) = 125\,°C.$$

It then follows that for the Carnot heat engine the thermal efficiency is:

$$\eta_{Carnot} = 1 - \frac{T_c}{T_{h,av}} = 1 - \frac{15 + 273}{125 + 273} = 0.276.$$

And, similarly for the endo-reversible heat engine:

$$\eta_{endo} = 1 - \sqrt{\frac{T_c}{T_{h,av}}} = 1 - \sqrt{\frac{15 + 273}{125 + 273}} = 0.149.$$

The power output for each heat engine then follows, since:

$$\dot{W} = \eta \dot{Q}_h.$$

So,

$$\dot{W}_{Carnot} = \eta_{Carnot} \dot{Q}_h = 0.276 \times 315 = 87.1\ kW;\ and$$
$$\dot{W}_{endo} = \eta_{endo} \dot{Q}_h = 0.276 \times 315 = 47.0\ kW.$$

Therefore, the prediction based on the endo-reversible assumption leads to 54% reduction in power output compared to the Carnot heat engine.

2.1.5   Work Potential: Exergy

In Section 2.1.2 a simple method for determining the amount of heat available within a waste-heat stream was described. Synonymous with this, it is also possible to determine the maximum amount of work that could be generated from a waste-heat stream at T_h if it was cooled down to the ambient temperature, T_0. The work potential, or exergy X, is given by

$$X = \int_{T_0}^{T_h} \left(1 - \frac{T_0}{T_h}\right) dQ, \tag{2.15}$$

where T_0 is the ambient, or dead-state temperature, and T_h is the heat-source temperature. For a waste-heat stream with a mass-flow rate $\dot{m}$ and constant specific-heat capacity c_p, the solution to Equation 2.15 leads to:

$$\dot{X} = \dot{m}c_p \left[(T_h - T_0) - T_0 \ln\left(\frac{T_h}{T_0}\right)\right], \tag{2.16}$$

whilst for a waste-heat stream with a variable specific-heat capacity c_p, the exergy is given as:

$$\dot{X} = \dot{m}[(h_h - h_0) - T_0(s_h - s_0)], \tag{2.17}$$

where h_h and s_h are the enthalpy and entropy of the waste-heat source, and h_0 and s_0 are the dead-state enthalpy and entropy of the heat-source fluid. It is noted that since we are dealing with a mass-flow rate, the exergy given by Equations 2.16 and 2.17 corresponds to the work rate or power.

In most processes in which heat is extracted to generate work, a heat stream will undergo a reduction in temperature. However, this temperature reduction may not necessarily be as large as that associated with being cooled all the way down to ambient temperature. Therefore, it is also convenient to define the work potential as the difference between the exergy contained within the inlet stream and that contained within the outlet stream:

$$\dot{X}_{\text{net}} = \dot{X}_{\text{in}} - \dot{X}_{\text{out}}, \tag{2.18}$$

where $\dot{X}_{\text{in}}$ and $\dot{X}_{\text{out}}$ are calculated using either Equation 2.16 or 2.17 depending on the fluid stream.

Example 2.2   For the same heat source and heat sink defined in Example 2.1, calculate the total work potential if the heat stream is cooled down to ambient temperature, and the net work potential for the defined heat-source temperature drop of 50 °C.

Solution
From Equation 2.16, the total exergy contained within the waste-heat stream can be calculated from:

$$\dot{X} = \dot{m}_{\text{h}} c_{p,\text{h}} \left[(T_{\text{hi}} - T_{\text{c}}) - T_{\text{c}} \ln \left(\frac{T_{\text{hi}}}{T_{\text{c}}} \right) \right]$$

$$\dot{X} = 1.5 \times 4.2 \times \left[(150 + 273) - (15 + 273) - (15 + 273) \ln \left(\frac{150 + 273}{15 + 273} \right) \right]$$

$$\dot{X} = 153.0 \text{ kW}.$$

To determine the net work potential for the defined temperature drop we can calculate the exergy still contained within the waste-heat stream at the outlet temperature.

$$\dot{X}_{\text{out}} = \dot{m}_{\text{h}} c_{p,\text{h}} \left[(T_{\text{ho}} - T_{\text{c}}) - T_{\text{c}} \ln \left(\frac{T_{\text{ho}}}{T_{\text{c}}} \right) \right]$$

$$\dot{X}_{\text{out}} = 1.5 \times 4.2 \times \left[(100 + 273) - (15 + 273) - (15 + 273) \ln \left(\frac{100 + 273}{15 + 273} \right) \right]$$

$$\dot{X}_{\text{out}} = 66.3 \text{ kW}.$$

Noting that inlet exergy is equal to the total value previously calculated (i.e. $\dot{X}_{\text{in}} = 153.02$ kW), it follows that:

$$\dot{X}_{\text{net}} = \dot{X}_{\text{in}} - \dot{X}_{\text{out}} = 153.0 - 66.3 = 86.7 \text{ kW}.$$

Therefore, for the heat source defined in Example 2.1, a maximum of 153 kW could be produced if the heat source was fully utilised. However, in a real process in which the heat-source is only cooled down to 100 °C, the maximum power that could be produced reduces to 86.7 kW, which corresponds to a 43% reduction. It is also worth noting the similarity between the net work potential and the power output predicted for the Carnot heat engine in Example 2.1. In other words, using the average heat-source temperature to calculate the Carnot efficiency gives us a very good estimate of the maximum power that can be obtained from a particular heat source.

2.1.6 Heat Pumps

A heat pump is a technology that absorbs heat from a low-temperature heat source, and by the addition of work to the system upgrades this heat and rejects it to a higher-temperature heat sink. It can therefore be thought of as the opposite of a heat engine. As for the heat engine, a heat pump also operates within a closed loop, and therefore the internal energy of the system at the start and end of the loop is equal. Therefore, from the first law of thermodynamics we have:

$$Q_c - Q_h = -W, \tag{2.19}$$

where Q_c is the heat absorbed by the system from the low-temperature heat source, Q_h is the heat rejected to the high-temperature heat sink and W is the work required to drive the heat pump.

The coefficient of performance (COP) ϕ of a heat pump is defined as the ratio of the heat generated by the heat pump to the work required to drive the cycle, therefore:

$$\phi = \frac{Q_h}{W} = \frac{Q_h}{Q_h - Q_c}. \tag{2.20}$$

Furthermore, if we assume the heat pump to be ideal, and therefore reversible, from the second law of thermodynamics it follows that:

$$\frac{Q_h}{T_h} = \frac{Q_c}{T_c}, \tag{2.21}$$

where T_h and T_c are the heat-sink and heat-source temperatures, respectively. Therefore:

$$\phi = \frac{T_h}{T_h - T_c}, \tag{2.22}$$

which is referred to as the Carnot COP of a heat pump, and this represents the theoretical maximum COP that can be obtained by a heat pump operating between a known heat-source temperature and heat-sink temperature.

However, in reality this COP will never be achieved by a heat pump owing to irreversibilities within the system. Synonymous with the heat engine, estimates for the endo-reversible COP of a heat pump have been investigated and include an empirical correlation suggested by Blanchard (1980):

$$\phi = \left(1 - \frac{T_c}{T_h + \Delta T_k}\right)^{-1}, \tag{2.23}$$

where ΔT_k is an empirical temperature difference, with values of 30 K being recommended. Another simple correlation was suggested by Velasco et al. (1997), and is given as:

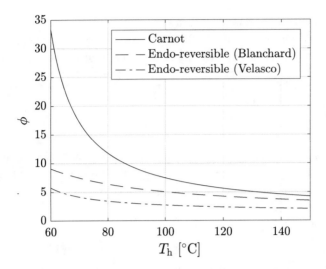

Figure 2.2 The effect of the heat-sink temperature T_h on the Carnot and endo-reversible coefficients of performance for a heat pump. The heat-source temperature is fixed at 50 °C

$$\phi = \sqrt{\frac{T_h}{T_h - T_c}}. \qquad (2.24)$$

The Carnot and endo-reversible COPs have been plotted as a function of the heat-sink temperature in Figure 2.2, assuming a fixed heat-source temperature of 50 °C. Ultimately, this figure is useful to provide a first-stage assessment of the expected COP for a heat pump based on a known heat-sink temperature. Furthermore, it also shows that as the temperature difference between the heat source and heat sink increases, the heat-pump COP reduces, which means an increasing amount of work is required to drive the system.

Example 2.3   A waste-heat stream of water at 45 °C, with a mass-flow rate of 1.2 kg/s and specific-heat capacity of 4.2 kJ/(kg K), is cooled down to 35 °C, and this heat is upgraded to a higher temperature using a heat pump. The heat pump rejects the upgraded heat to a heat sink which is maintained at a constant temperature of 70 °C. Estimate the heat produced and the power required to drive the heat pump, for a Carnot heat pump, and for a more realistic endo-reversible heat pump.

Solution
Firstly, calculate the amount of heat that is transferred into the heat pump:

$$\dot{Q}_c = \dot{m}_c c_{p,c}(T_{ci} - T_{co})$$
$$\dot{Q}_h = 1.2 \times 4.2 \times (45 - 35) = 50.4 \text{ kW}.$$

The COP of a Carnot or endo-reversible heat pump can be calculated from the heat-source and heat-sink temperatures. However, as in Example 2.1, the heat source undergoes a temperature reduction and is not constant. Again, a reasonable approach is to use the average heat-source temperature.

$$T_{c,av} = 0.5(T_{ci} + T_{co}) = 0.5(45 + 35) = 40\,°C.$$

It then follows that for a Carnot heat pump, the COP can be calculated:

$$\phi_{Carnot} = \frac{T_h}{T_h - T_{c,av}} = \frac{70 + 273}{70 - 40} = 11.43.$$

And for an endo-reversible heat pump, either Equation 2.23 or 2.24 can be used:

$$\phi_{Blanchard} = \left(1 - \frac{T_{c,av}}{T_h + \Delta T_k}\right)^{-1} = \left(1 - \frac{40 + 273}{(70 + 273) + 30}\right)^{-1} = 6.22$$

$$\phi_{Velasco} = \sqrt{\frac{T_h}{T_h - T_c}} = \sqrt{\frac{70 + 273}{70 - 40}} = 3.38.$$

From Equation 2.20, it follows that:

$$\dot{Q}_h = \frac{\phi \dot{Q}_c}{\phi - 1},$$

and hence:

$$\dot{Q}_{h,Carnot} = \frac{\phi_{Carnot}\dot{Q}_c}{\phi_{Carnot} - 1} = \frac{11.43 \times 50.4}{11.43 - 1} = 55.23\,kW$$

$$\dot{Q}_{h,Blanchard} = \frac{\phi_{Blanchard}\dot{Q}_c}{\phi_{Blanchard} - 1} = \frac{6.22 \times 50.4}{6.22 - 1} = 60.06\,kW$$

$$\dot{Q}_{h,Velasco} = \frac{\phi_{Velasco}\dot{Q}_c}{\phi_{Velasco} - 1} = \frac{3.38 \times 50.4}{3.38 - 1} = 71.56\,kW.$$

Finally, the power output for each heat pump can be calculated:

$$\dot{W}_{Carnot} = \frac{\dot{Q}_{h,Carnot}}{\phi_{Carnot}} = \frac{55.23}{11.43} = 4.83\,kW$$

$$\dot{W}_{Blanchard} = \frac{\dot{Q}_{h,Blanchard}}{\phi_{Blanchard}} = \frac{60.06}{6.22} = 9.66\,kW$$

$$\dot{W}_{Velasco} = \frac{\dot{Q}_{h,Velasco}}{\phi_{Velasco}} = \frac{71.56}{3.38} = 21.16\,kW.$$

From these results it is clear that Blanchard and Velasco models predict that the heat pump would generate 8.7% and 29.6% more heat than the Carnot heat pump, but would require 200% and 438% more power, respectively. In other words, a real heat pump requires significantly more power to generate one unit of heat than a Carnot heat pump. The large difference between the two endo-reversible models also highlights a difficulty in obtaining precise estimates for heat pump performance. These are, however, useful to provide a ball-park figure that can be refined through further study.

2.1.7   Absorption Chillers

An absorption chiller is a technology that absorbs heat from a high-temperature heat source and uses this heat to provide a cooling load to a low-temperature heat sink. Denoting the heat addition from the high-temperature heat source as Q_h, and the cooling load as Q_c, the COP of an absorption chiller is given as:

$$\psi = \frac{Q_c}{Q_h}, \qquad (2.25)$$

and represents the amount of cooling generated for each unit of heat that is input into the system for the high-temperature heat source. The precise operating principle of an absorption chiller will be discussed in more detail in Chapter 3, but essentially consists of four heat-transfer processes. Two of these processes are the ones that have already been discussed, and they correspond to the system absorbing heat from the high-temperature heat source, and from the low-temperature heat sink, respectively. The other two processes are heat-rejection processes in which the system rejects heat to an ambient heat sink. A simple way to imagine an absorption chiller is to decouple it into a heat engine and a heat pump. The heat engine operates between the heat source and ambient heat sink, and the work generated by this heat engine drives the heat pump which transfers heat from the cold sink to the ambient heat sink. Therefore, ψ is the product of the heat-engine thermal efficiency η and the heat-pump COP. Finally, it should also be noted that the COP of a refrigeration heat pump and a heat-upgrade heat pump are related by $\phi_{\text{heating}} = \phi_{\text{cooling}} + 1$, hence:

$$\psi = \frac{Q_c}{Q_h} = \frac{W}{Q_h}\frac{Q_c}{W} = \eta(\phi - 1). \qquad (2.26)$$

Therefore, for a Carnot heat engine operating with a Carnot heat pump, the Carnot COP for an absorption chiller is given as:

$$\psi = \left(1 - \frac{T_0}{T_h}\right)\left(\frac{T_c}{T_0 - T_c}\right), \qquad (2.27)$$

where T_h, T_c and T_0 are the temperatures of the high-temperature heat source, low-temperature heat sink and ambient heat sink, respectively. Similarly, the COP for an endo-reversible absorption chiller is given by:

$$\psi = \left(1 - \sqrt{\frac{T_0}{T_h}}\right)\left(\left(1 - \frac{T_c}{T_0 + \Delta T_k}\right)^{-1} - 1\right), \qquad (2.28)$$

when using the Blanchard (1980) correlation to model the heat pump, and:

$$\psi = \left(1 - \sqrt{\frac{T_0}{T_h}}\right)\left(\sqrt{\frac{T_0}{T_0 - T_c}} - 1\right), \qquad (2.29)$$

when using the Velasco et al. (1997) correlation to model the heat pump.

The Carnot and endo-reversible COPs have been plotted as a function of the heat-sink temperature and heat-source temperature in Figure 2.3, assuming a fixed ambient

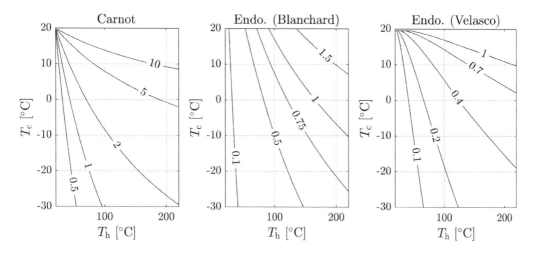

Figure 2.3 The effect of the heat-source temperature T_h and heat-sink temperature T_c on the Carnot and endo-reversible COP for an absorption chiller. The ambient heat-sink temperature is fixed at 20 °C

heat-sink temperature of 20 °C. Ultimately, this figure is useful to provide a first-stage assessment of the expected COP for an absorption chiller based on a known heat-source temperature and heat-sink temperature.

Example 2.4 A waste-heat stream of pressurised water at 200 °C, with a mass-flow rate of 0.8 kg/s and specific-heat capacity of 4.2 kJ/(kg K), is cooled down to 175 °C, and this heat is used to provide cooling using a chiller. The chiller is used to maintain the cold heat sink at a constant temperature of 5 °C, whilst the ambient temperature is 25 °C. Estimate the amount of cooling that can be produced for a Carnot chiller and for a more realistic endo-reversible chiller.

Solution
Firstly, calculate the amount of heat that is transferred into the system:

$$\dot{Q}_h = \dot{m}_h c_{p,h} (T_{hi} - T_{ho})$$
$$\dot{Q}_h = 0.8 \times 4.2 \times (200 - 175) = 84 \text{ kW}.$$

As per Examples 2.1 and 2.3, the average heat-source temperature is used to calculate the COP for the system:

$$T_{h,av} = 0.5(T_{hi} + T_{ho}) = 0.5(200 + 175) = 187.5 \text{ °C}.$$

The COP for heat chiller system can then be calculated from Equations 2.27, 2.28 and 2.29, respectively:

$$\psi_{\text{Carnot}} = \left(1 - \frac{T_0}{T_{h,av}}\right)\left(\frac{T_c}{T_0 - T_c}\right) = \left(1 - \frac{25 + 273}{187.5 + 273}\right)\left(\frac{5 + 273}{25 - 5}\right) = 4.90$$

$$\psi_{\text{Blanchard}} = \left(1 - \sqrt{\frac{T_0}{T_{h,av}}}\right)\left(\left(1 - \frac{T_c}{T_0 + \Delta T_k}\right)^{-1} - 1\right)$$

$$= \left(1 - \sqrt{\frac{25 + 273}{187.5 + 273}}\right)\left(\left(1 - \frac{5 + 273}{(25 + 273) + 30}\right)^{-1} - 1\right) = 1.09$$

$$\psi_{\text{Velasco}} = \left(1 - \sqrt{\frac{T_0}{T_{h,av}}}\right)\left(\sqrt{\frac{T_0}{T_0 - T_c}} - 1\right)$$

$$= \left(1 - \sqrt{\frac{25 + 273}{187.5 + 273}}\right)\left(\sqrt{\frac{25 + 273}{25 - 5}} - 1\right) = 0.56.$$

The cooling generated by each system is then as follows:

$$\dot{Q}_{c,\text{Carnot}} = \psi_{\text{Carnot}}\dot{Q}_h = 4.90 \times 84 = 412\,\text{kW}$$

$$\dot{Q}_{c,\text{Blanchard}} = \psi_{\text{Blanchard}}\dot{Q}_h = 1.09 \times 84 = 91.3\,\text{kW}$$

$$\dot{Q}_{c,\text{Velasco}} = \psi_{\text{Velasco}}\dot{Q}_h = 0.56 \times 84 = 47.0\,\text{kW}.$$

These results show that when considering a real chiller, the cooling power that can be generated is significantly lower than that predicted for the Carnot chiller; more precisely, the estimates based on the Blanchard and Velasco heat pump correlations predict cooling powers that are 22% and 11% of the Carnot calculation. In other words, a real chiller can be expected to produce significantly less cooling from a given heat source than a Carnot chiller. As for the heat pump, large difference between the two endo-reversible models also highlights a difficulty in obtaining precise estimates for chiller performance. However, as stated previously, these are useful to provide a ball-park figure that can be refined through further study.

2.2   Heat-Exchanger Fundamentals

Although the focus of this book is on technologies that convert waste heat into other forms of energy, such as electricity or cooling, heat exchangers have an important role to play. In the first instance, heat exchangers can be used to directly recover waste heat for the purpose of heating a different process stream. If there is a direct demand for heat on-site, direct heat recovery in this way will always be the most suitable option, owing to its simplicity and effectiveness. In the second instance, heat exchangers are required to interface a waste-heat stream to a waste-heat recovery technology, transferring the available waste heat into the system.

2.2.1   Energy Balance

The purpose of a heat exchanger within waste-heat recovery is typically to transfer heat from a high-temperature heat source to a colder process stream. In this process, the heat-source undergoes a reduction in temperature, whilst the process stream undergoes an increase in temperature. Applying an energy balance to a heat exchanger, the amount of heat transferred $\dot{Q}$ is:

$$\dot{Q} = \dot{m}_h c_{p,h}(T_{hi} - T_{ho}) = \dot{m}_c c_{p,c}(T_{co} - T_{ci}), \tag{2.30}$$

where $\dot{m}_h$ and $\dot{m}_c$ are the mass flow rates of the hot and cold fluids, $c_{p,h}$ and $c_{p,c}$ are the specific-heat capacities of the hot and cold fluids, T_{hi} and T_{ci} are the inlet temperatures of the hot and cold fluids and T_{ho} and T_{co} are the outlet temperatures of the hot and cold fluids.

2.2.2 Heat-Exchanger Effectiveness

The maximum heat transfer is possible when the stream with the lowest heat-capacity rate (i.e. $(\dot{m}c_p)_{min}$) undergoes the maximum possible change in temperature, which for a counter-flow heat exchanger is the difference between the two inlet temperatures, hence:

$$\dot{Q}_{max} = (\dot{m}c_p)_{min}(T_{hi} - T_{ci}), \tag{2.31}$$

where

$$(\dot{m}c_p)_{min} = \min\{(\dot{m}c_p)_h, (\dot{m}c_p)_c\}. \tag{2.32}$$

The heat-exchanger effectiveness is defined as the ratio of the actual heat transferred to the maximum. Therefore, combining Equations 2.30 and 2.31, the heat-exchanger effectiveness is given as:

$$\varepsilon = \frac{\dot{Q}}{\dot{Q}_{max}} = \frac{(\dot{m}c_p)_h(T_{hi} - T_{ho})}{\dot{Q}_{max}} = \frac{(\dot{m}c_p)_c(T_{co} - T_{ci})}{\dot{Q}_{max}}. \tag{2.33}$$

2.2.3 Heat-Exchanger Sizing

From a thermodynamic point of view, maximising ε will always result in the maximum recovery of heat from a waste-heat stream. However, a higher heat-exchanger effectiveness will always correspond to a heat exchanger with a larger heat-transfer area requirement. The required heat-transfer area can be calculated using the following:

$$\dot{Q} = U A \, \Delta T_{log}, \tag{2.34}$$

where U is the overall heat-transfer coefficient (with units W/(m^2 K)), A is the heat-exchanger area (with units m^2) and ΔT_{log} is the log-mean temperature difference, which for a counter-flow heat exchanger is given as:

$$\Delta T_{log} = \frac{(T_{hi} - T_{co}) - (T_{ho} - T_{ci})}{\log\left(\frac{T_{hi} - T_{co}}{T_{ho} - T_{ci}}\right)}. \tag{2.35}$$

To investigate this trade-off between thermodynamic performance and the required heat-exchanger area, a simple case study has been considered in which hot water

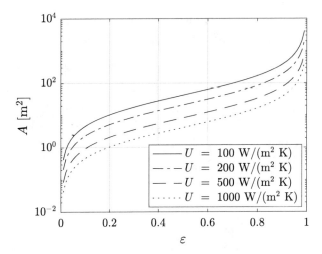

Figure 2.4 The effect of the heat-exchanger effectiveness ε on the required heat-transfer area A at different overall heat-transfer coefficients U

at 100 °C is used to heat cold water at 15 °C. Both are assumed to have a heat-capacity rate of $\dot{m}c_p = 4{,}200$ W/K. For different overall heat-transfer coefficients, the heat-transfer area can be calculated as a function of ε, and the results are shown in Figure 2.4. Clearly, a very high effectiveness corresponds to a very large heat-transfer area requirement. Furthermore, as the overall heat-transfer coefficient increases, the required area reduces.

2.2.4   Cost Considerations

From the previous section it is clear that a trade-off between maximising the heat-exchanger effectiveness and the required heat-transfer area exists. To covert the required heat-transfer area into a corresponding cost, the following correlation taken from Smith (2005) is applied:

$$C = C_b \left(\frac{A}{A_b} \right)^{0.68}, \tag{2.36}$$

where C is the cost of the heat exchanger, and C_b and A_b are the base size and cost terms, and are given by Smith (2005) as $\$3.28 \times 10^4$ and 80 m², respectively. The costs associated with the areas shown in Figure 2.4 are given in Figure 2.5, and again, a clear correlation between increasing heat-exchanger cost and increasing heat-exchanger effectiveness is observed.

Finally, the thermodynamic performance and cost of the heat exchanger can be coupled together by considering the specific investment cost (SIC) of the heat exchanger, defined here as the total cost divided by the amount of heat recovered (i.e. SIC = $C/\dot{Q}$), with units \$/kW. The results for this case study are shown in Figure 2.6. The results from this plot are interesting in that they suggest there is an optimal effectiveness at

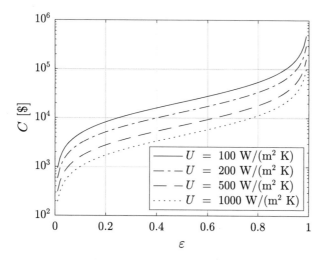

Figure 2.5 The effect of the heat-exchanger effectiveness ε on the cost of the heat exchanger C at different overall heat-transfer coefficients U

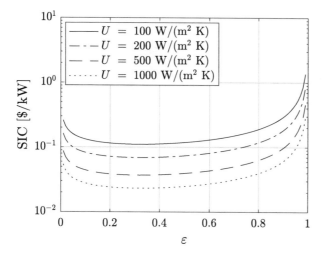

Figure 2.6 The effect of the heat-exchanger effectiveness ε on the SIC of the heat exchanger at different overall heat-transfer coefficients U

which the SIC is minimised. For example, for this case study in particular, this appears somewhere between $\varepsilon = 0.2$ and $\varepsilon = 0.4$.

Overall, this brief overview of heat-exchanger fundamentals has demonstrated an important trade-off between maximising the thermodynamic performance whilst minimising the cost of a waste-heat-recovery heat exchanger. The precise optimal heat-exchanger design for a particular waste-heat stream will depend on many factors such as the temperature, mass flow rate and the medium of the heat source and the process fluid, in addition to available space, available capital or other economic

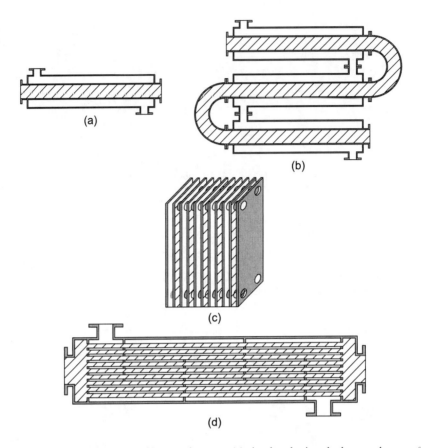

Figure 2.7 Different types of heat exchangers: (a) simple tube-in-tube heat exchanger; (b) stacked tube-in-tube heat exchanger; (c) plate heat exchanger; and (d) shell-and-tube heat exchanger

factors. None-theless, the analysis presented is suitable for a first-stage assessment of installing a direct heat-recovery technology.

2.2.5   Types of Heat-Exchanger

The design and geometry of a heat exchanger will be dependent on the temperature, mass flow rate and the medium of the heat source and the process fluid. In general, the three most common types of heat exchangers are tube-and-tube, shell-and-tube and plate heat exchangers, and these are shown in Figure 2.7.

A tube-and-tube heat exchanger consists of two concentric tubes with different diameters. One fluid passes through the inner tube, whilst the second fluid passes through the annular space, typically in the opposite direction. By comparison, the shell-and-tube design consists of multiple smaller diameter tubes placed within a large shell. One fluid enters at one side of the heat exchanger and passes through the small tubes, whilst the second fluid occupies the shell and passes in the opposite direction. Furthermore, baffles are placed within the heat exchanger, which act to direct the shell-side fluid perpendicularly across the tube bundle. The main advantage of the shell-and-tube design over the tube-and-tube design is the greater heat-transfer area and a longer contact time between the two fluids, leading to enhanced heat transfer and more effective heat recovery. Both the tube-and-tube and shell-and-tube heat geometries can be further improved by using finned tubes, which further enhance the heat-transfer area. Finally, a plate heat exchanger is constructed from multiple thin plates which are positioned together with a small separation gap. The two fluids flow in opposing directions and occupy the flow channels alternatively. A large number of plates, and a small separation gap, ensure a high heat-transfer area for a small fluid flow rate. Much like the other heat exchanger designs, heat transfer can be enhanced by the addition of chevrons to the plates.

Although heat exchangers can be classified by their geometry, they can also be classified according to their intended duty. The following list details a number of specific heat exchangers that are relevant within waste-heat recovery applications.

- An *economizer* recovers heat from low-temperature exhaust gases for the purpose of heating a liquid. Typically, the cold liquid passes through a finned-tube bundle over which the hot exhaust gases pass.
- A *recuperator* recovers heat from exhaust gases and uses this heat to preheat another gas stream. Recuperators typically resemble a plate heat exchanger in their construction. A common application is within gas turbines where the hot turbine exhaust is used to preheat the compressor exhaust before combustion.
- A *regenerator* is typically used to recycle heat within an industrial process. Firstly, the hot fluid passes over a thermal storage material that absorbs heat from the heat stream. The cold fluid is then heated by passing it over the thermal store at a later stage.
- A *boiler* uses hot exhaust gases to convert a fluid from liquid to gas. One such example would be a shell-and-tube heat exchanger in which the fluid to be heated occupies the shell, whilst the hot exhaust gases pass through the tubes.

2.3 Plant Audit

In order to evaluate the feasibility of installing waste-heat recovery technologies at a particular industrial plant, an audit is required in the first instance. This should identify all available waste-heat steams and quantify these in terms of their thermodynamic parameters, in addition to considering the practical aspects of installing a waste-heat recovery technology.

2.3.1 Thermodynamic Considerations

A plant audit should quantify all available waste-heat sources in terms of their temperature, mass-flow rate, specific-heat capacity and composition. From this, meaningful parameters, such as the heat potential and work potential, can be calculated using the methods described in Section 2.1.2 and 2.1.5.

2.3.2 Practical Considerations

Alongside the thermodynamic considerations, the plant audit must also address a number of additional factors, which can be summarised as follows:

- *Heat source medium.* Typical waste-heat sources include subcooled liquids, hot air, exhaust gas, pressurised water, waste water and steam. The waste-heat stream will influence the design of the heat exchanger used to recover the waste heat.
- *Composition.* The composition of exhaust gases can vary, depending on the industrial process, but often can include CO_2, NO_x, sulphides and water. The composition will influence material selection, and can also limit the heat-source outlet temperature to avoid condensate and the deposit of corrosive substances on the heat exchanger walls.
- *Interaction with existing plant.* A waste-heat recovery technology should not interfere with the original industrial process. This applies additional constraints, such as the maximum allowable pressure drop, on the waste-heat recovery system. Furthermore, the location of the heat source and the space available for the installation of a waste-heat recovery technology need to be carefully considered. Finally, it may also be necessary to interface multiple heat sources to the same waste-heat recovery system, and the design of such a system needs to be carefully planned.
- *Dynamics.* Many heat streams from industrial processes can be transient in nature. Therefore the variation in the waste-heat temperature and mass flow rate should be evaluated, and the effect on the waste-heat recovery potential determined. It may be necessary to install an intermediate heat-transfer loop which can act as a buffer between the fluctuating heat source and the waste-heat recovery system. Alternatively, it may be necessary to interface the waste-heat recovery technology with a thermal storage technology.
- *Plant energy demand.* Finally, an audit of the plant's energy consumption should be conducted. This should establish the on-site demand for power, heating and cooling, in addition to determining economic factors such as the cost of electricity and other primary fuels. Furthermore, the load factor of both the waste-heat source and the energy demand should be determined to ensure the annual operating hours are sufficient to amortize the capital costs of the waste-heat recovery system.

3 Heat Conversion and Upgrading Technologies

The aim of this chapter is to introduce the reader to the main technologies available for the conversion of waste heat into other forms of energy such as power, heating or cooling. These three applications correspond to three different groups of technologies, namely heat engines, heat pumps and chillers. Each group of technologies will be discussed in terms of operating principles, operational limits and the current commercial status. Furthermore, general selection criteria will be discussed.

3.1   Overview of Technologies

The conversion of waste heat into heating, cooling and power can be achieved by a variety of different technologies. An overview of the technologies available is presented in Figure 3.1, in which technologies are also divided into passive and active. A passive technology is defined here as a technology that uses a waste-heat stream to heat up a colder process stream. This could be achieved directly (e.g., recirculating a hot exhaust gas into a production process) either through the installation of a heat exchanger to transfer the heat to a different process stream or through thermal-energy storage in which the heat is stored until it is required at a later time. By comparison, active technologies are defined as technologies that utilise waste heat for the purpose of generating other forms of energy, and these technologies are the focus of this chapter. Primarily this concerns heat engines, which are used to generate electricity from a waste-heat stream; heat pumps, which are used to upgrade waste heat to a higher temperature; and finally chillers, which are capable of using waste heat to provide cooling.

The selection of a suitable technology for a particular waste-heat stream will depend on many factors, including the heat available; the demand for heating, cooling or power on-site; and the cost of the system and the economic incentives for installing a technology. Nonetheless, particular technologies from Figure 3.1 are at the forefront of waste-heat recovery; for example, organic Rankine cycles (ORCs), vapour-compression heat pumps (VCHPs) and absorption chillers. Within the literature, a number of authors have attempted to evaluate and compare different technologies. For example, Tauveron, Colasson and Gruss (2014, 2015) categorised different waste heat to power systems based on heat-source temperature and power output; van de Bor, Infante Ferreira and Kiss (2015) compared VCHPs and heat engines

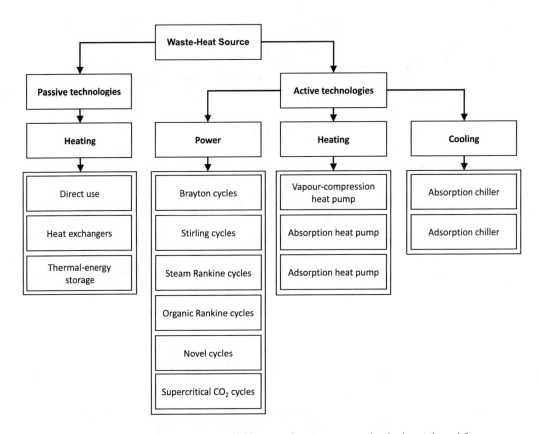

Figure 3.1 A summary of the available waste-heat recovery technologies. Adapted from Brückner et al. (2015)

for low temperature waste-heat recovery below 100°C; and Brückner et al. (2015) compared vapour-compression and absorption heat pumps with absorption chiller systems. There have also been attempts to rank available waste-heat streams and match these to different technologies (Oluleye, Jobson & Smith 2015; Oluleye et al. 2016). However, given that studies such as these typically relate to site-specific boundary conditions, it remains difficult to make general recommendations. Thus, given the number of technologies identified in Figure 3.1, it is apparent that when faced with the decision of which technology to install, a large number of technologies must be considered. This requires either certain technologies to be ruled out from the start or a detailed assessment of each technology to be conducted. The aim of this chapter is to outline and discuss the various technologies available and evaluate current commercial systems in terms of thermodynamic and economic performance metrics, enabling general recommendations to be made.

3.2   Heat Engines

The first group of technologies to be discussed recover waste heat and convert this heat into mechanical power, which can then be used directly or converted into electricity, which is either used on-site or exported to the national electricity grid. A number of different heat engines can be considered and these are discussed in the following subsections.

3.2.1   Brayton Cycle

The Brayton cycle is widely used to generate power from the combustion of natural gas and involves the compression of air from ambient conditions, an isobaric heat addition and then an expansion across a turbine back down to ambient pressure. The efficiency of this simple Brayton cycle can be vastly improved by installing a recuperator. Furthermore, closing the cycle by installing a heat-rejection heat-exchanger between the turbine and compressor can facilitate the use of alternative working fluids such as helium, argon or carbon dioxide. However, with the exception of supercritical CO_2 (sCO_2) applications, which will be discussed briefly later in this chapter, Brayton cycles are generally not suitable for utilising heat at the low-temperature levels associated with waste-heat recovery applications.

3.2.2   Stirling Cycle

In the Stirling cycle, the cyclic change in volume caused when a piston moves up and down in a cylinder is used to compress and expand the working fluid to produce power. Between the compression and expansion stroke, the working fluid temperature increases due to an isochoric (constant volume) heat addition process, whilst the working fluid undergoes an isochoric heat rejection process during the return stroke before the cycle can repeat. For low-temperature applications, including solar applications and waste-heat recovery, Stirling engines have potential advantages such as simple construction and reliability, which make them cost-effective and energy-efficient solutions (Kongtragool & Wongwises 2003; Wang et al. 2016). Alongside conventional Stirling engines, there are also novel heat engines based on the Stirling engine concept, which include thermo-acoustic heat engines (Backhaus & Swift 2002; Tijani & Spoelstra 2011), free-piston expanders (Walker & Senft 1985) and liquid piston Stirling engines. This latter group is also referred to as thermo-fluidic oscillators and includes both single-phase engines (Stammers 1979) and two-phase engines (Smith 2012; Kirmse et al. 2017). However, at present further research is required to improve performance whilst demonstrating the potential of these heat engines for industrial-scale applications.

3.2.3   Steam Rankine Cycle

The steam Rankine cycle consists of four main components, namely a pump, evaporator, expander and condenser, and a schematic of the system is shown in Figure 3.2. Saturated or sub-cooled water is pressurised in the pump (1 to 2) before passing through the evaporator where it absorbs heat from the heat source (2 to 3). This process preheats, fully evaporates and then superheats the water to generate steam.

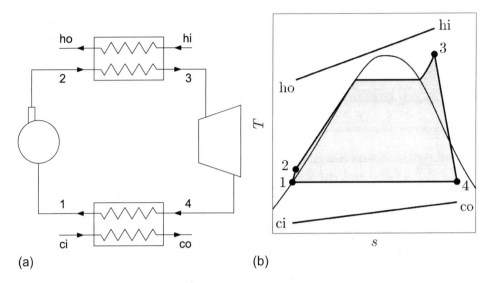

Figure 3.2 The operating principle of a Rankine cycle described in terms of the system components (a) and the thermodynamic cycle represented on a temperature–entropy diagram (b)

The high-temperature, high-pressure steam then expands across the turbine producing mechanical work (3 to 4), which can be converted into electricity using a generator. The steam then passes through the condenser, rejecting heat to a heat sink (4 to 1). During this process the steam is pre-cooled to the condensation temperature and then fully condensed before the cycle can repeat.

Although the steam Rankine cycle can be an efficient thermodynamic cycle, its performance deteriorates as the heat-source temperature drops below around 300 °C. Moreover, water has a high condensation temperature at ambient pressure, which either has a detrimental effect on the cycle performance or requires a sub-atmospheric condenser. This second option increases system complexity since the condenser must be designed to prevent air ingress, and increases costs since low densities result in large heat exchangers. Furthermore, due to the shape of the saturated vapour dome of water, sufficient superheating is required to prevent condensation within the turbine. Nonetheless, despite these challenges, steam Rankine cycle systems may be suitable candidates for high-temperature waste-heat recovery applications under certain conditions.

3.2.4   Organic Rankine Cycle

A natural progression from the steam Rankine cycle is to consider the use of alternative working fluids, and this has led to the ORC. Although there has been a significant interest in ORC technology in recent decades, the idea of using alternative fluids is not new. In a review paper, Colonna et al. (2015) accredited T. Howard as first patenting the idea in 1826. However, it is only recently with increasing concerns regarding climate change and the need for new sustainable and efficient power systems that the interest in these systems has increased. An ORC system is exactly the same as the steam Rankine cycle with the same components and system layout. However, an organic fluid facilitates lower-temperature heat sources, typically between 80 and 400 °C, to be converted into power more efficiently than steam. This makes them a prime candidate for waste-heat recovery applications within this temperature range. A detailed discussion of ORC technology is provided in Chapter 5, as such only a brief overview of some of the key aspects will be given here.

The plethora of potential working fluids means that working-fluid selection remains an active area of research. Generally speaking, a relationship between the critical temperature of the fluid and heat source temperature can be observed, with fluids with low critical temperatures being favoured for relatively low-temperature heat sources and vice versa. Thus, it is possible, in a general fashion, to classify working fluids in this manner (Chen, Goswami & Stefanakos 2010; White & Sayma 2018), and it follows that the optimal working fluid will depend on the available heat source. In terms of working-fluid selection, an ideal working fluid should have preferential thermodynamic behaviour, be environmentally friendly, chemically stable, non-flammable, non-toxic and cheap. The criteria for selecting a working fluid are summarised within a number of studies that can be found within the literature, although Badr, Probert and O'Callaghan (1985) were among the first to outline these considerations in detail, with the notable exclusion of environmental properties. Typically the selection of an optimal working fluid results from identifying a group of potential fluids that meet all the necessary legislative and safety requirements and then comparing each fluid in terms of both thermodynamic performance and overall system cost. However, it is worth noting that, despite the large number of working fluids available, there are only really a handful of fluids that are commonly used within commercial ORC power systems (Colonna et al. 2015).

Besides fluid selection and cycle design, a significant area of research is the development of small-scale ORC systems for waste-heat recovery applications, alongside solar and combined heat and power. In particular, waste-heat recovery from automotive and stationary engines is a large area of current research. One challenge facing these small-scale systems is the development of suitable expanders, with positive-displacement expanders such as scroll, screw and reciprocating-piston machines being under investigation, alongside turbo-expanders.

3.2.5 Variations on the Rankine Cycle

The basic steam and organic Rankine cycles can be further adapted to enhance thermodynamic performance. These cycles are the recuperated, working-fluid mixture, supercritical and trilateral or partially evaporated, cycles.

In the recuperated cycle an additional heat exchanger is installed within the system and this heat exchanger uses the hot expander exhaust gas to preheat the working fluid before it enters the evaporator. A schematic of a recuperated Rankine cycle is shown in Figure 3.3. The heat transfer that occurs as a result of installing a recuperator is given

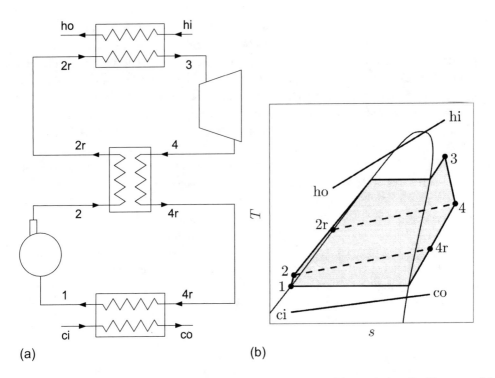

Figure 3.3 The operating principle of a recuperated Rankine cycle described in terms of the system components (a) and the thermodynamic cycle represented on a temperature–entropy diagram (b)

by the dashed lines and corresponds to the hot exhaust gas cooling down (4 to 4r) and the cold liquid being heated up (2 to 2r). The installation of a recuperator reduces the amount of heat required to produce the same power output, and hence increases the cycle efficiency, but is also associated with a more complex, and therefore costly system. It is also worth noting that in some waste-heat recovery applications, where one is more interested in maximising the power generated from waste-heat source, it may not be necessary to include a recuperator as a recuperator actually restricts how much heat can be extracted from the heat source. On the other hand, there are waste-heat recovery applications where the minimum heat-source outlet temperature is restricted due to a downstream process or to prevent condensation of the exhaust gases that could cause corrosion. Therefore the benefits of recuperation should be considered carefully for the particular waste-heat recovery application in question.

The remaining three cycles do not involve additional components but improve the thermodynamic performance of a basic Rankine cycle by obtaining a better thermal match between the working fluid and the heat source. The advantage of an improved thermal match is to reduce irreversibility with the system, particularly within the heat-addition process, thus improving the efficiency of the heat conversion process and resulting in a better thermodynamic performance. The three cycles are a Rankine cycle

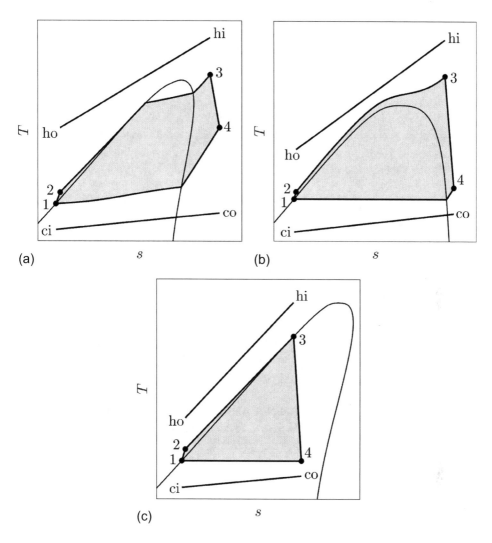

Figure 3.4 Alternative thermodynamic cycles based on the Rankine cycle. Clockwise from top-left corner: working-fluid mixture, supercritical and trilateral cycles

operating with a mixture, a supercritical cycle and the trilateral, or partially evaporated cycle. These three cycles are shown in Figure 3.4.

Using a working-fluid mixture achieves a better thermal match between the heat-source and heat-sink streams since a mixture undergoes non-isothermal phase change. This means that the temperature of the mixture increases during evaporation and decreases during condensation. As an example of the effect this could have on cycle's performance, Lecompte et al. (2014) report possible improvements in the second law efficiency between 7 and 14% for low-temperature applications (120 to 160 °C). However, using a mixture can lead to reduced heat-transfer coefficients compared to pure fluids, which is likely to lead to larger heat-transfer area requirements (Azzolin, Bortolin & Del Col 2016). Although slightly different to the cycle shown in

Figure 3.4(a), another type of thermodynamic cycle that uses a working-fluid mixture is the Kalina cycle. In the Kalina cycle a mixture of ammonia and water is used that vaporises at different rates when heat is applied (Kalina 1984). Controlling the concentration of the two fluids and splitting the fluid into ammonia-rich and ammonia-poor streams allows flexibility to optimise the system to maximise the thermodynamic performance. However, compared to a basic Rankine cycle operating with a mixture, the Kalina cycle is more complicated owing to the need to separate and combine the ammonia-rich and ammonia-poor streams at different points within the cycle.

In the supercritical cycle, the two-phase evaporation is removed entirely by pressurising the working fluid above its critical pressure. Therefore, the heat-addition process does not involve any two-phase heat transfer and could lead to improvements in performance compared to the conventional ORC system. However, it is not only necessary to understand that supercritical cycles correspond to higher operating pressures, requiring more expensive pipework and pumps and expanders suitable for high-pressure operation, it is also necessary that we improve our understanding of heat-transfer phenomena under supercritical pressures.

The final cycle, the trilateral cycle, avoids two-phase evaporation by expanding the working fluid from a saturated-liquid state. Therefore the heat addition is only used to preheat the working fluid from the sub-cooled pumpoutlet condition to a saturated liquid. If the working fluid is expanded from a two-phase state (i.e. with a vapour quality between 0 and 1), the cycle may be termed a partially evaporated cycle. The trilateral cycle has been explored in detail in the works by (Smith 1993; Smith & da Silva 1994; Smith, Stosic & Aldis 1996). However, the major challenge facing these systems is the development and widespread commercialisation of a suitable expansion machine that can tolerate two-phase conditions.

Ultimately, these alternative cycles can improve the thermodynamic performance of the system, but introduce challenges to the system design that can lead to increased costs. For example, for all three of these systems, the reduced temperature difference between the working fluid and heat source is advantageous as it reduces irreversibility within the cycle but leads to lower rates of heat transfer (i.e., $Q = UA\Delta T_{log}$). As a consequence the required heat-transfer area increases which can increase the cost of the system. As such, it becomes important to consider the trade-off between the thermodynamic performance benefits and cost. In summary, Lecompte et al. (2015) concluded that whilst some of these cycles are promising from a thermodynamic viewpoint, the economic viability of these systems is not yet clear. Furthermore, there remain technical challenges such as a lack of suitable components and a general lack of experimental validation for these novel systems. These challenges need to be addressed before technologies based on these cycles can be commercialised.

3.2.6   Supercritical CO$_2$ Cycles

Another technology that has received a significant amount of attention in the past few years are power systems that operate with sCO$_2$ as the working fluid. Much of the early work on sCO$_2$ cycles was completed in the 1960s by Feher (1968) and Angelino

(1968), but interest was recently reignited by Dostal (2004). Supercritical CO_2 cycles can be considered for a wide range of heat-source temperatures, which could range from a few hundred degrees Celsius right up to operating temperatures comparable to air Brayton cycles. The main applications that sCO_2 cycles are currently being considered for are as a replacement for steam in nuclear power systems, concentrated solar power plants, alongside waste-heat recovery applications. Compared to steam, operating with sCO_2 has significant benefits including high thermal efficiencies at relatively moderate heat-source temperatures and a significantly smaller power block, owing to the high operating pressures and the corresponding high density. Thus, in theory the use of sCO_2 could lead to more efficient and cost-effective power systems than those employed today. However, at present, sCO_2 technology is still in its infancy, and, with a few exceptions, sCO_2 systems are not yet commercially available. Currently, there exist significant challenges around the design of these systems, particularly around material selection for the challenging design environment, alongside the design of suitable turbomachines and heat exchangers for these systems. A detailed overview on the current status of sCO_2 technology can be found in Brun, Friedman and Dennis (2017).

Depending on the application, a sCO_2 cycle could either take the form of a Brayton cycle or a Rankine cycle (Figure 3.5). In almost all cases, a recuperator is included within these cycles to improve thermal efficiency, and depending on the size of the system and the cycle, recompression and reheating may also be considered. Operating a sCO_2 Rankine cycle does, in general, offer an improvement in thermal efficiency compared to the Brayton cycle, owing to a lower compression work. However, operating a Rankine is dependent upon being able to condense CO_2 at a temperature below 31.1 °C, which is the critical temperature of CO_2 above which there is no longer a distinction between the liquid and vapour phases. Thus, to realistically achieve a Rankine cycle a heat sink with a temperature in the region of around 10 °C is likely to be required, which is unfeasible for many applications.

3.2.7 Thermodynamic Performance of Heat Engines

Figures 3.6 and 3.7 show typical theoretical and practical thermal efficiencies for a number of power-generation systems, including Brayton, steam Rankine and organic Rankine cycle power plants over a range of heat-source temperatures. These efficiency values are also compared to the maximum theoretical reversible (Carnot) and endo-reversible efficiencies defined by Equations 2.13 and 2.14, respectively. These results clearly demonstrate the potential of ORC systems at low temperatures, and the suitability of steam Rankine systems as the maximum operating temperature increases. Furthermore, it is clear that at low temperatures there is no thermodynamic benefit in operating a Brayton cycle.

3.2.8 Current Commercial Status

From Figures 3.6 and 3.7 it is clear that steam and Rankine cycles are the most suitable heat engines for the conversion of waste heat into power, and therefore it follows

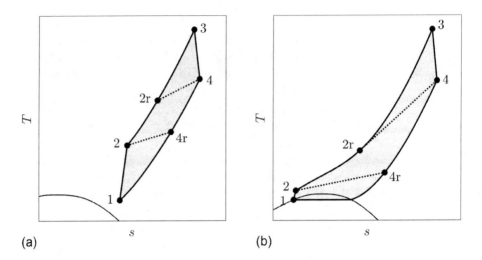

Figure 3.5 Supercritical CO_2 Brayton cycle (a) and Rankine cycle (b)

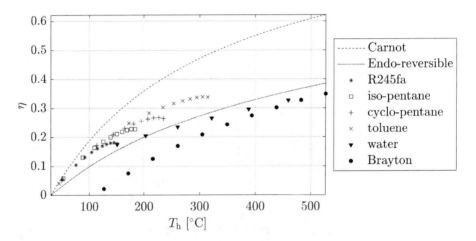

Figure 3.6 Thermal efficiency of common thermodynamic heat engines over a range of heat-source temperatures. Results are based on a simple cycle analysis of a Brayton and Rankine cycle with a fixed heat-sink temperature of 30 °C

that these technologies are the dominant systems available within the market. A summary of information collated from publicly available data sheets from manufacturers of these two systems is given in Tables 3.1 and 3.2 and includes the power output from the system, heat-source temperatures required to drive the system and the expander technology used within the system. The commercially available steam systems listed have been restricted to power outputs below 10 MW, and with the exception of the Heliex Power system, are designed for heat-source temperatures up to around 500 °C. The Heliex Power system is designed for wet steam expansion and is therefore suitable for lower temperature applications.

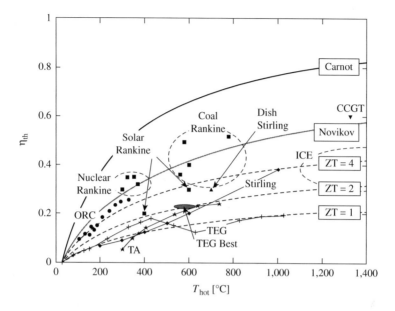

Figure 3.7 Thermal efficiency of common thermodynamic heat engines and thermoelectric generators (TEGs) over a range of heat-source temperatures. Circles represent actual ORC and Kalina cycle plants; squares are for various steam Rankine systems; triangles for solar dish Stirling and combined-cycle gas turbines (CCGTs); diamonds for conventional Stirling and stars for thermo-acoustic (TA) engines. Original figure is taken from Markides (2015)

Table 3.1 Summary of commercially available steam Rankine cycle systems with power outputs below 10 MW. Data collated from data sheets available within the open literature.

	$\dot{W}$ [kW]	T_{hi} [°C]	Expander
Elliott	> 50	< 482	Axial
Heliex Power	70–500	150–300	Screw
MAN	1,000–10,000	250–450	Axial
MAN	4,000–40,000	< 520	Axial
Siemens	> 75	< 530	Axial
Spilling	100–5,000	< 480	Piston, axial
VOITH	40–360	300–400	Piston

References for these companies are listed in Appendix A

The systems listed in Tables 3.1 and 3.2 help to establish the current operational limits of commercial steam and organic Rankine cycle systems in terms of required heat-source temperatures. Most of the steam Rankine cycle systems are quoted as being suitable for operation with heat-source temperatures between 300 and 500 °C. In comparison, ORC systems are more suited to lower heat-source temperatures, typically between 100 and 350 °C. Therefore there appears to be a transition from steam to organic fluids as the heat-source temperature reduces below 350 °C, and below 300 °C ORC systems become the dominant technology. Alongside the three major

Table 3.2 Summary of commercially available ORC systems. Data collated from data sheets available within the open literature.

	$\dot{W}$ [kW]	T_{hi} [°C]	Expander
Atlas Copco	2,000–45,000	250–650	Radial
BEP	55–800	85–150	Screw
Calnetix	125	130	Radial
DeVeTec	50–270	250–350	Piston
Electratherm	35–110	77–122	Screw
Enertime	100–5,000	90–200	Turbine
Enogia	5–100	80–200	Turbine
Exergy	100–50,000	90–300	Radial-out
GE	23,000–43,000	450–550	Radial
Maxxtec	300–2,400	320	Turbine
ORMAT	250–20,000	130–220	Axial
Siemens	400–1,500	300	Axial
Triogen	95–165	350–530	Radial
Turboden	200–300	200–310	Axial
Turboden	600–6,800	270–315	Axial

References for these companies are listed in Appendix A

ORC manufacturers, there are a number of companies offering alternative ORC systems. Furthermore, within these systems a range of different expander technologies are used, although axial and radial turbines are typically favoured for high-power and high-temperature applications, whilst volumetric expanders can be considered when the heat-source temperature and power output reduce.

The data provided in Tables 3.1 and 3.2 is useful to assess the commercially available heat-to-power technologies but gives no information regarding the cost of these systems. To assess cost, it is useful to refer to the specific investment cost (SIC), which is typically defined as the capital cost per kilowatt of power rating. Within the literature, there have been multiple attempts to assess SICs for ORC technology. However, since data is scarce, or not publicly available, for real installations these studies should not be taken as absolute, but should instead be interpreted as being representative of general trends. For example, a relationship between increasing system size and reducing SIC is observed, with values generally ranging between 10,000 €/kW for a 10 kW system and 1,000 €/kW as the system size exceeds a few MW (Quoilin et al. 2013). More specifically, Elson, Hampson and Tidball (2015) conducted a market assessment of waste-heat recovery technologies and estimated the SIC and operational and maintenance costs of steam and organic Rankine cycles (Table 3.3). Alternatively, Lemmens (2016) collated data on ORC plants and plotted the plant size against SIC in €/kW for a range of applications (N.B.: $/€ = 1.1) (Figure 3.8).

3.3   Heat Pumps and Chillers

The second group of technologies are heat pumps, which transfer heat between two reservoirs at two different temperatures. Depending on the application and the mode

Table 3.3 Estimated specific investment costs in $/kW and operational and maintenance costs (O&M) in $/kWh for steam Rankine cycle and ORC technologies.

		Costs for different capacities				
Technology	Cost type	50–500 kW	0.5–1 MW	1–5 MW	5–20 MW	>20 MW
Steam	SIC	$3,000	$2,500	$1,800	$1,500	$1,200
	O&M	$0.013	$0.009	$0.008	$0.006	$0.005
ORC	SIC	$4,500	$4,000	$3,000	$2,500	$2,100
	O&M	$0.020	$0.015	$0.013	$0.012	$0.010

Reproduced from Elson, Hampson & Tidball (2015)

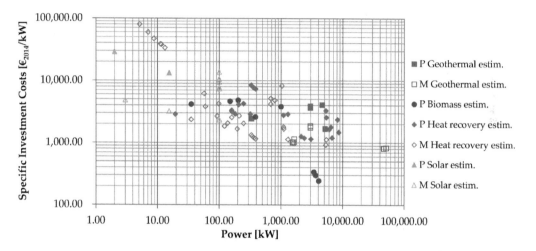

Figure 3.8 Estimated costs of ORC projects (P) and modules (M) reported in the literature, in 2014. Original figure is taken from Lemmens (2016)

of operation, a heat pump can be used to provide either a heating or cooling load. In the case of waste-heat recovery, this corresponds to either upgrading a heat source to a higher temperature or using a heat source to provide cooling.

3.3.1 Vapour-Compression Heat Pumps

The VCHP is the most common and technically mature heat pump technology. The basic VCHP consists of a compressor, condenser, expansion valve and evaporator, which are coupled together to form a closed-loop system (Figure 3.9). The working fluid is initially compressed from either a saturated vapour or superheated state (1 to 2) before being sub-cooled and condensed within the condenser (2 to 3). The working fluid then expands across the expansion valve (3 to 4) before being re-evaporated in the evaporator (4 to 1). During the condensation process, the working fluid rejects heat to the high-temperature heat sink, whilst in the evaporation process, the working fluid absorbs heat from the low-temperature heat source. In this way, the heat pump

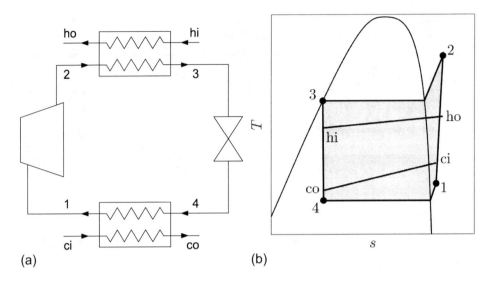

(a)   (b)

Figure 3.9 The operating principle of a vapour-compression heat pump is described in terms of the system components (a) and the thermodynamic cycle represented on a temperature–entropy diagram (b)

provides both a heating effect to the heat sink and a cooling effect to the heat source, and can therefore be used for both heating and cooling applications.

The commercialisation of VCHPs began in the 1950s, although it wasn't until the 1970s that the uptake of the technology began to significantly increase (Hepbasli & Kalinci 2009). Today, typical applications include ground-source heat pumps, air-source heat pumps and air-conditioning systems. However, with increasing concerns over climate change and depleting fuel reserves, heat pumps offer a practical solution for improving energy efficiency since they can recirculate environmental and waste heat back into heat-production processes within domestic, commercial and industrial applications. Much like an ORC, a VCHP can operate with a range of different working fluids, and ultimately the selection criteria for a working fluid are not that dissimilar to those outlined for ORCs. Up until the mid-1990s, chlorofluorocarbons were commonly used due to their preferential thermodynamic performance. However, as environmental regulations have tightened it has become necessary to consider alternative working fluids such as hydrocarbons and hydrofluorocarbons, with the more recent trend towards hydrofluoroolefins which have a low global warming potential.

Alongside research into new working fluids, another focus has been to improve the energy efficiency of VPHC systems. In particular, this relates to improved system efficiencies in addition to achieving higher temperature lifts that allow the production of heat at higher temperatures. Generally speaking, high temperature lifts require high compression ratios that cannot be achieved over a single-stage compression. Therefore, multi-stage compression in which multiple compressors are installed in series, or cascaded systems in which a topping and bottoming cycle are coupled together, can be considered (Chua, Chou & Yang 2010). Another possibility is to operate a transcritical heat pump, where the working fluid is compressed to a pressure greater than its critical pressure, thus removing the isothermal condensation process. Ultimately, as suggested by Ommen et al. (2015), different heat-pump systems are suited to different heat production temperatures and temperature lifts, and thus the choice of cycle and fluid will be application dependent.

In general, the drive towards high-temperature lift heat pumps is particularly relevant to industrial waste-heat recovery applications. Jakobs, Cibis and Laue (2010) discuss industrial heat-pump applications and state there are many industrial areas in which process water between 40 and 90 °C and steam between 100 and 200 °C are required, and high-temperature heat pumps could easily supply this heat using the local environment, cooling water steams or waste heat. In general, it is suggested that high-temperature heat pumps could supply heat up to around 120 to 150 °C (Jakobs, Cibis & Laue 2010; Brückner et al. 2015; van de Bor, Infante Ferreira & Kiss 2015).

3.3.2   Absorption and Adsorption Heat Pumps and Chillers

An absorption heat pump is different to a VCHP in that it does not require a compressor to drive the cycle, but instead exploits the properties of a working-fluid mixture comprising a refrigerant and an absorbent. The system is primarily constructed from a generator, condenser, evaporator and an absorber, and a schematic of the system is shown in Figure 3.10. Initially, the refrigerant is fully absorbed within the absorbent and this mixture enters the generator via the pump. Then as a high-temperature heat source is applied to the generator, the refrigerant evaporates from the absorbent, thus separating the mixture into a refrigerant vapour and an absorbent liquid. The refrigerant vapour passes through the condenser where it condenses and rejects heat to ambient surroundings, before passing through an expansion valve. The low-pressure refrigerant liquid then passes through the evaporator where it evaporates, thus absorbing heat from a low-temperature heat source and providing a cooling effect. Finally, the refrigerant vapour moves into the absorber. Meanwhile, the absorbent liquid that is generated in the generator passes through another expansion valve and also get collected in the absorber. Then, within the absorber, heat is rejected to another ambient heat sink, during which the refrigerant re-condenses and is reabsorbed by the absorbent. A pump then recirculates the mixture back into the generator where the cycle can repeat. The two most common working-fluid pairs are water/liquid bromide and ammonia/water.

Typically, within waste-heat recovery, an absorption heat pump is used to provide cooling from a waste-heat stream, in which it is typically referred to as an absorption chiller. As with the other technologies discussed, investigating new working-fluid mixtures and more advanced cycle architectures remains an active area of research. The challenge is finding an optimal refrigerant-absorbent mixture that meets thermodynamic and environmental requirements whilst developing absorption heat pumps with higher coefficients of performance, or absorption heat pumps that are capable of achieving high temperature lifts. A comprehensive review of different cycle architectures was completed by Srikhirin, Aphornratana and Chungpaibulpatana (2001)

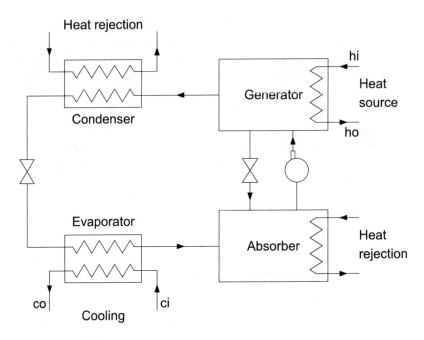

Figure 3.10 Schematic of an absorption heat pump/absorption chiller

alongside a comprehensive comparison of absorption systems. In general, an absorption chiller system could provide cooling from waste-heat streams as high as 200 °C, with coefficients of performance ranging between 0.2 and 1.7 (Srikhirin, Aphornratana & Chungpaibulpatana 2001; Brückner et al. 2015). An adsorption chiller can be used for the same effect, and operates in a very similar way to an absorption chiller. However, it differs that instead of the refrigerant dissolving into another fluid, the fluid adsorps onto the surface of a solid and then during heating, it desorps from the surface. However, Chan, Ling-Chin and Roskilly (2013) state that absorption chiller systems are the most developed, and most suitable technology to provide cooling from a waste-heat stream.

3.3.3   Current Commercial Status

Both VCHP and AC systems are commercially available for the conversion of heat into heating and cooling, respectively. Following a brief survey of the market, the authors were able to collate data on these systems, and the results from this review are shown in Figures 3.11 and 3.12.

In Figure 3.11, the commercially available VCHP systems are described by the heat-source inlet temperature, and the temperature lift achieved by the heat pump ΔT_{lift}, which is defined as:

$$\Delta T_{\text{sink}} = T_{\text{ho}} - T_{\text{ci}}, \tag{3.1}$$

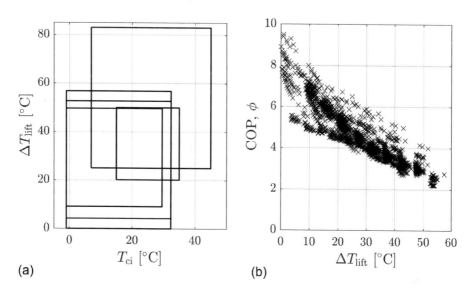

Figure 3.11 Map of current commercially available heat pumps in terms of heat-source inlet temperature T_{ci} and temperature lift ΔT_{lift} (a), and in terms of ΔT_{lift} and COP ϕ (b). Data collected from the heat-pump manufacturers listed in Appendix A

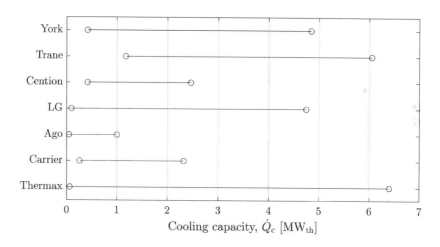

Figure 3.12 Summary of commercially available absorption chillers in terms of their cooling capacity. Data collected from absorption chiller manufacturers listed in Appendix A

where T_{ho} is the outlet temperature of the high-temperature heat sink, and T_{ci} is the inlet temperature of the low-temperature heat source.

From Figure 3.11 it is apparent that the commercially available VCHP systems are capable of upgrading waste heat at temperatures below 50 °C, up to a maximum of 90 °C. Therefore, at their current commercial status the technology is suited for upgrading very low-temperature waste-heat streams, such as cooling water, to a higher temperature. The heating capacity of these systems is found to range between 5 kW$_{th}$ and 3.4 MW$_{th}$, whilst COPs range between 2 and 6 for temperature lifts between 20 and 60 °C.

In terms of commercial AC systems, these systems are not necessarily designed for waste-heat recovery applications, but instead can be driven by fossil fuels, saturated steam or hot water. Typically, application data is supplied for the AC system when it is being driven by either hot water at 90 °C or saturated steam at 0.8 MPa (170 °C), and the system is used to provide cold water at 7 °C. Quoted COPs are generally between 0.5 and 1.5, although the COP of the system will increase as the temperature difference between the heat source and heat sink increases. The cooling capacity of these systems ranges between 64 kW$_{th}$ and 6.4 MW$_{th}$.

In a similar fashion to the commercial Rankine cycle systems, the SICs for VCHP and AC systems are also dependent on the size of the system and the temperature difference between the heat source and heat sink. However, it is possible to make some general comments on SICs based on reports available within the literature. More specifically, the SIC of a VCHP system could be between 150 and 500 $/kW$_{th}$ (Johnson & Choate 2008; Element Energy 2014; Brückner et al. 2015), whilst the SIC of an AC system could range between 2,500 $/kW$_{th}$ for systems below 10 kW$_{th}$ and 400 €/kW$_{th}$ for systems approaching 300 kW$_{th}$. Extrapolating these results for larger systems, it could be assumed that the SIC for larger systems could reach a few hundred € /kW$_{th}$ (Eicker & Pietruschka 2009).

3.4   Novel Technologies and Future Developments

In addition to conventional thermodynamic cycles, there are also a number of emerging technologies in the research and development stage that can generate electricity directly from heat, and in the future these could be viable technologies for power generation from waste-heat sources. These technologies include thermoelectric, piezoelectric, thermionic and thermo-photovoltaic devices. Several of these have undergone prototype testing in automotive applications and are under development for industrial waste-heat recovery applications. However, at present the efficiencies of these devices do not exceed the efficiencies of conventional thermodynamic cycles, and these devices are associated with very high SICs, quoted as in excess of 20,000 $/kW (Johnson & Choate 2008).

3.5   Summary and Focus of This Book

From the review of technologies presented in this chapter, three technologies are identified as the most suitable for waste-heat recovery applications. These are the Rankine cycles (operating with either steam or an organic fluid), VCHPs and absoprtion chillers, which can be used for power, heating and cooling applications, respectively. A summary of these technologies is provided in Table 3.4.

Table 3.4 Summary of commercially available WHR technologies

Technology	Application	T_{hi}	Size	Cost
Heat engines	Power	$> 100\,°C$	$> 100\,kW$	$1,000–10,000\,€/kW$
Heat pumps	Heating	$< 90\,°C$	$5\,kW_{th}–4\,MW_{th}$	$150–500\,€/kW$
Chillers	Cooling	$< 200\,°C$	$60\,kW_{th}–7\,MW_{th}$	$400–2,500\,€/kW$

From Table 3.4 it is possible to make the following very general recommendations regarding technology selection at different heat-source temperature levels:

- For low-temperature heat sources below $100\,°C$, heat pumps and absorption chillers are the most suitable technology.
- Between 100 and $200\,°C$, both heat engines and absorption chillers could be considered and the decision would be primarily be based on whether there is a demand for cooling or electricity on-site.
- For temperatures above $200\,°C$, heat engines are the most appropriate technology.
- Unlike heat pumps and absorption chillers, the electricity generated from a heat engine can be easily exported to the grid. Therefore, heat engines are the most widely applicable technology for heat source temperatures exceeding $100\,°C$.

Ultimately, although these recommendations may be useful, the selection of a waste-heat recovery technology for a specific waste-heat stream will be dependent on many factors, such as the amount of heat available, the on-site power, heating or cooling demand and economic factors such as the cost of electricity or natural gas, operating hours and government incentives. Therefore, the correct selection can only be made after completing an evaluation of each technology.

4 Technology-Agnostic Modelling

In this chapter, technology-agnostic models for heat engines and heat pumps will be developed and applied to waste-heat recovery applications. A technology-agnostic model is defined here as a model of a heat engine or heat pump that considers only the heat-source and heat-sink conditions, whilst the inner workings of the system (i.e. operating conditions, working fluid, component performance, etc.) are ignored. These technology-agnostic models can be extremely useful during a preliminary assessment of the waste-heat recovery potential at a particular site.

4.1 Heat Engines

4.1.1 Thermodynamic Modelling

The Carnot efficiency of a heat engine operating between a heat source and a heat sink at fixed temperatures was given by Equation 2.13, and is repeated here for clarity as:

$$\eta = 1 - \frac{T_c}{T_h},$$
(4.1)

where T_h and T_c are the temperatures of the heat source and heat sink, respectively. This efficiency is the maximum thermal efficiency that a heat engine can achieve. However, in reality this efficiency can never be achieved due to irreversibilities within the system. Novikov (1958) derived a remarkably simple relationship for the optimal thermal efficiency of a heat engine, and this is referred to as the endo-reversible efficiency:

$$\eta = 1 - \sqrt{\frac{T_c}{T_h}},$$
(4.2)

which was also previously defined by Equation 2.14.

Within the derivation of Equations 4.1 and 4.2, the heat source and heat sink are both assumed to be infinite such that they remain at a fixed temperature. However, within a practical heat engine, the heat source and the heat sink can no longer be assumed to be isothermal since they undergo a temperature reduction and temperature increase, respectively. To account for this behaviour, the full heat engine can be thought of as many differential heat engines each operating between different heat-source and heat-sink temperatures, as shown in Figure 4.1. Assuming each differential

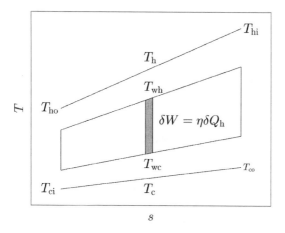

Figure 4.1 Finite-time heat engine operating between a non-isothermal heat source and heat sink represented on a T–s diagram

engine is an ideal cycle, the second law of thermodynamics states that $dQ/T =$ constant. Noting that $dQ = \dot{m}c_p dT$, one obtains:

$$\int_{T_h}^{T_{hi}} \frac{(\dot{m}c_p)_h dT_h}{T_h} = \int_{T_{ci}}^{T_c} \frac{(\dot{m}c_p)_c dT_c}{T_c} , \qquad (4.3)$$

where $\dot{m}$ is the mass-flow rate, c_p is the specific-heat capacity and T_{hi} and T_{ci} are the heat-source and heat-sink inlet temperatures, respectively. Equation 4.3 can be solved to obtain an expression for T_c as a function of T_h, which can be substituted into either Equation 4.1 or 4.2 to obtain the efficiency as a function of T_h only. Then the work generated by a heat engine in which the heat source is cooled to a temperature T_{ho} is:

$$\dot{W} = \int_{T_{ho}}^{T_{hi}} \eta dQ_h = \dot{m}_h \int_{T_{ho}}^{T_{hi}} c_{p,h}(T_h)\eta(T_h)dT_h . \qquad (4.4)$$

If the temperature dependence of the specific-heat capacity is removed, Equations 4.3 and 4.4 can be solved analytically. In this case, the relationship between T_c and T_h is given as follows:

$$T_c = T_{ci}\left(\frac{T_{hi}}{T_h}\right)^{\frac{1}{\alpha}} , \qquad (4.5)$$

where

$$\alpha = \frac{(\dot{m}c_p)_c}{(\dot{m}c_p)_h} . \qquad (4.6)$$

Then, if we consider a Carnot heat engine as an example, the work generated is given by:

$$\dot{W} = (\dot{m}c_p)_h \left[(T_{hi} - T_{ho}) - \alpha T_{ci}\left(\left(\frac{T_{hi}}{T_{ho}}\right)^{\frac{1}{\alpha}} - 1\right) \right] , \qquad (4.7)$$

and the heat input and thermal efficiency are given by:

$$\dot{Q} = (\dot{m}c_p)_h(T_{hi} - T_{ho}), \text{ and} \tag{4.8}$$

$$\eta = \frac{\dot{W}}{\dot{Q}} = 1 - \frac{\alpha T_{si,i}\left(\left(\frac{T_{so,i}}{T_{so,o}}\right)^{\frac{1}{\alpha}} - 1\right)}{T_{so,i} - T_{so,o}}. \tag{4.9}$$

Finally, from Equation 4.7 it can be seen that an optimal heat-source outlet temperature must exist at which optimal power is produced by the heat engine. First consider the case in which $T_{ho} = T_{hi}$. Since there is no change in the heat-source temperature, the heat input and therefore work produced by the heat engine is zero. Similarly, considering the case where $\alpha = 1$ and where the heat source is cooled right down to the heat-sink inlet temperature (i.e. $T_{ho} = T_{ci}$) implies a heat exchange between the heat source and heat sink with an effectiveness of 1, and $T_{co} = T_{hi}$. Clearly, in this case the work produced by the heat engine is also equal to zero. Therefore, an optimum must exist and will be obtained at the condition $d\dot{W}/dT_{ho} = 0$. The analysis presented can also be extended to consider an endo-reversible heat engine (i.e. Equation 4.2), or to a case when the heat-source and heat-sink specific-heat capacities vary with temperature. In this case, an algebraic solution may not be possible, but nonetheless the system of equations can be solved numerically.

In order to complete an economic assessment of a heat engine, it is necessary to be able to predict the temperature differences within the heat engine. Within such a heat engine, there is a temperature difference between the heat source and working fluid, denoted ΔT_h, and a temperature difference between the heat sink and the working fluid, denoted ΔT_c, and these can also be seen in Figure 4.1. The thermal efficiency given by Equation 4.2 is the endo-reversible thermal efficiency of the heat engine and therefore already accounts for these temperature differences. Therefore, if the endo-reversible heat engine is thought of as being equivalent to a Carnot heat engine operating between a lower heat-source temperature, denoted T_{wh}, and a higher heat-sink temperature, denoted T_{wc}, it follows that:

$$1 - \sqrt{\frac{T_c}{T_h}} = 1 - \frac{T_{wc}}{T_{wh}} = 1 - \frac{T_c + \Delta T_c}{T_h - \Delta T_h}. \tag{4.10}$$

Furthermore, the heat transfer from the heat source to the working fluid is given as:

$$\dot{Q}_h = k_h(T_h - T_{wh}), \tag{4.11}$$

and the heat transfer from the working fluid to the heat sink is calculated as:

$$\dot{Q}_c = k_c(T_{wc} - T_c), \tag{4.12}$$

where k is the heat conductance, which is also equivalent to the product of the overall product of the overall heat-transfer coefficient U and the heat-exchanger area A. Introducing $\beta = k_c/k_h$ and solving Equations 4.10, 4.11 and 4.12, the following expressions for T_{wh} and T_{wc} are obtained:

$$T_{\text{wh}} = \frac{T_{\text{h}} + \beta\sqrt{T_{\text{h}}T_{\text{c}}}}{\beta + 1}, \tag{4.13}$$

$$T_{\text{wc}} = \frac{\sqrt{T_{\text{h}}T_{\text{c}}} + \beta T_{\text{c}}}{\beta + 1}, \tag{4.14}$$

and hence:

$$\Delta T_{\text{h}} = T_{\text{h}} - T_{\text{wh}} = \left(\frac{\beta}{\beta + 1}\right)(T_{\text{h}} - \sqrt{T_{\text{h}}T_{\text{c}}}), \tag{4.15}$$

$$\Delta T_{\text{c}} = T_{\text{wc}} - T_{\text{c}} = \frac{\sqrt{T_{\text{h}}T_{\text{c}}} - T_{\text{c}}}{\beta + 1}. \tag{4.16}$$

Therefore, for a heat engine operating between specified heat-source and heat-sink temperatures, the temperature differences between the heat source and the working fluid, and between the heat sink and the working fluid can be obtained if the heat-conductance ratio β is known.

4.1.2 Thermodynamic Model Validation

In order to validate the technology-agnostic heat-engine thermodynamic model, a brief review of commercially available technologies has been completed. Data sheets were collected for a number of commercially available heat engines that are based on the organic Rankine cycle. From these the required heat-source and heat-sink conditions were extracted and the performance evaluated using the technology-agnostic heat-engine model. This could then be compared to the thermal efficiency quoted directly by the manufacturer. The results are presented in Figure 4.2. Before discussing this figure, the following points should be noted:

- Where the heat-sink temperature was not available, $T_{\text{ci}} = 25\,°\text{C}$ was assumed;
- ORC manufacturers are only included for which efficiency and/or heat-source temperatures are specified within the available literature;
- Enogia and Electratherm are relatively small-scale systems, hence the lower efficiencies compared to other systems;
- Triogen systems typically operate with high heat-source temperatures, and are designed for CHP applications; hence, thermal efficiency may be sacrificed in favour of overall efficiency.

Referring to Figure 4.2, it should be noted that there are two endo-reversible lines: 'fixed' and 'var. source'. The 'fixed' line corresponds to an isothermal heat source and heat sink, and therefore corresponds to the maximum efficiency for an endo-reversible heat engine. However, since the heat source and heat sink are both isothermal, this case also corresponds to $\dot{W} = 0$. In comparison, the 'var. source' line corresponds to a heat engine where an isothermal heat sink is assumed but the heat source is cooled right down to the heat-sink temperature. Under this condition the heat engine will generate the maximum power but will also correspond to the lowest thermal efficiency. In

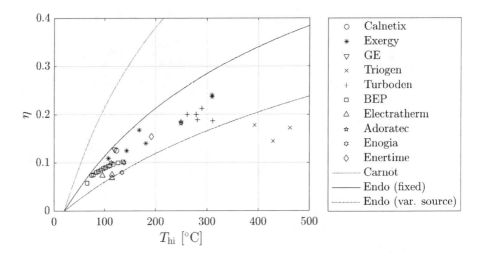

Figure 4.2 Comparison between manufacturer-quoted thermal efficiencies and efficiencies calculated using the technology-agnostic heat-engine model. The fixed source endo-reversible line corresponds to fixed heat-source and heat-sink temperatures, the variable source line corresponds to a varying heat-source temperature but with a fixed heat sink, and the variable source and sink line corresponds to a non-isothermal heat source and a non-isothermal heat sink

this manner, these two cases can be considered as reasonable extremes corresponding to the maximum and minimum thermal efficiency one would expect. Referring to the manufacturer-quoted efficiencies, it is observed that the majority of the real ORC systems lie between these two limiting cases, which provides a reasonable validation of the model. To verify the model further, the manufacturer data points should be compared to the more detailed heat-engine model that accounts for a variable heat sink. However, given the lack of more detailed information on the heat-source and heat-sink conditions for each real ORC system, such a comparison is not currently possible.

4.1.3   Economic Model

The economic benefit of installing a waste-heat recovery heat engine arises from the energy savings obtained, which means it is no longer necessary to purchase either electricity or natural gas from an external source. For a heat engine the annual savings S in \$ is given by:

$$S = n\dot{W}(C_e - C_{o\&m}),\qquad(4.17)$$

where n is the number of operating hours per annum, $\dot{W}$ is the heat-engine work output in kW, C_e is the cost of electricity in \$/kWh and $C_{o\&m}$ are the associated operating and maintenance costs, also given in \$/kWh.

The economic performance of a heat engine can then be evaluated by the specific investment cost (SIC) in \$/kW, the payback period (PB) in years, the net present value

(NPV) in \$ or the levelised cost of electricity (LCOE) in \$/kWh. These are given by the following four equations, respectively:

$$\text{SIC} = \frac{C_0}{\dot{W}} \, ; \tag{4.18}$$

$$\text{PB} = \frac{C_0}{S} \, ; \tag{4.19}$$

$$\text{NPV} = -C_0 + \sum_{t=1}^{t_{\max}} \frac{S}{(1+r)^t} \, ; \tag{4.20}$$

$$\text{LCOE} = \frac{C_0 + \sum\limits_{t=1}^{t_{\max}} \frac{\dot{W} n C_{\text{o\&m}}}{(1+r)^t}}{\sum\limits_{t=1}^{t_{\max}} \frac{\dot{W} n}{(1+r)^t}} \, , \tag{4.21}$$

where C_0 is the total investment cost, r is the discount rate, t is the current year and $t_{\max}$ is the operational lifespan.

Clearly, to determine all of the economic performance indicators it is necessary to estimate the total investment cost C_0. The simplest way to determine C_0 is to interpolate cost data for existing installations to estimate the cost of the technology. However, different technologies that do the same job can often have different costs associated with them, which means this approach is not suitable for a technology-agnostic approach. Instead, C_0 is approximated using cost functions for the different components. For the heat engine this means having cost functions for the four main components, namely: the pump, evaporator, expander and condenser. The total investment cost is then simply the summation of each component cost.

The costs of an expander (in £) and pump (in \$), denoted C_{ex} and C_{pu}, respectively, are estimated using the cost functions given by:

$$C_{\text{ex}} = \exp(a_1 + a_2 \log \dot{W}_{\text{ex}}), \tag{4.22}$$

and

$$C_{\text{pu}} = b_1 \left(\frac{\dot{W}_{\text{pu}}}{4} \right)^{b_2} , \tag{4.23}$$

respectively. These are taken from Seider, Seader and Lewin (2009) and Smith (2005), and the following values are recommended: $a_1 = 7.32$, $a_2 = 0.81$, $b_1 = 9.84 \times 10^3$, $b_2 = 0.55$. It should be noted that Equation 4.23 requires the pump work to be known, which is not estimated by the agnostic heat-engine model. However, for a simple estimate of the pump size, a back work ratio of 5% is assumed (i.e. the pump consumes 5% of the expander work).

The costs of the evaporator and condenser are dependent upon the temperature differences between the heat source and the working fluid, and the heat sink and the working fluid, respectively, and these have been defined by Equations 4.15 and 4.16. Considering that in a real heat engine the heat source and heat sink are non-isothermal the temperature differences at the heat-exchanger inlets and outlets will be different.

Table 4.1 Constants for Equation 4.26, taken from Hewitt (1994)

	c_1	c_2	c_3
Hot exhaust gas	1,855	−0.8729	0.4961
Cold air	3,960	−0.8368	0.3776
Cold water	2,424	−0.9361	0.0802

In this case, the log-mean temperature difference $\Delta T_{\log}$ is given by:

$$\Delta T_{\log} = \frac{\Delta T_i - \Delta T_o}{\ln\left(\frac{\Delta T_i}{\Delta T_o}\right)}, \tag{4.24}$$

where the subscripts 'i' and 'o' refer to the temperature differences at the inlet and outlet of the evaporator or condenser, respectively. This then leads to the calculation of the product of the overall heat-transfer coefficient U and the heat exchanger area A (Equation 4.25):

$$UA = \frac{Q}{\Delta T_{\log}}, \tag{4.25}$$

The cost of the heat exchanger (in £) can then be estimated using the cost function published by Hewitt (1994):

$$C = c_1(UA)^{c_2} + c_3. \tag{4.26}$$

The values for c_1, c_2 and c_3 depend upon the heat-source or heat-sink fluid, in addition to the type of heat exchanger selected. By example, values are provided in Table 4.1 for a waste-heat recovery heat engine using shell-and-tube heat exchangers. The heat source is assumed to be a hot exhaust gas, whilst the heat sink is either cooling water or air.

Having calculated the cost of each component, it is then necessary to convert each component cost to the current market cost using the Chemical Engineering Plant Cost Index, and convert the component costs into the correct currency.

4.1.4   Economic Model Validation

In addition to validating the thermodynamic model, the cost functions defined previously have been validated against estimated costs for ORC waste-heat recovery systems that were gathered by Lemmens (2016). This comparison is shown in Figure 4.3, and is based on a heat-sink temperature of $T_{ci} = 25\,°C$, a heat-capacity ratio of $\alpha = 5$ and a heat-conductance ratio of $\beta = 1$. Unfortunately, the temperature levels are not defined in Lemmens (2016), but it seems reasonable to assume that most ORC waste-heat recovery plants will be operating with heat-source temperatures between 150 °C and 250 °C. Therefore, it is easily observed that in general the heat-engine costs predicted using the defined cost functions agree well with the data available from

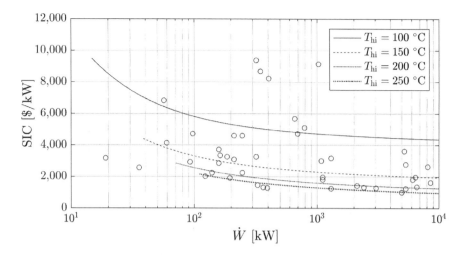

Figure 4.3 Specific investment costs for a different size heat engines operating under different heat source temperatures. The circles correspond to data points taken from Lemmens (2016), whilst the solid lines correspond to results from the technology-agnostic model. Technology-agnostic results obtained for an air-cooled heat engine at $T_{ci} = 25\,°\mathrm{C}$, with $c_{p,c} = 4200$ J/(kg K), $\alpha = 5$ and $\beta = 1$

the literature. It is also interesting to comment on the general trends observed in Figure 4.3. Clearly, the SIC is strongly dependent on the heat-source temperature and the plant size, with large, high-temperature systems corresponding to the most favourable SIC, and small, low-temperature system corresponding to the least favourable SIC.

4.1.5   Application of the Model

Having now developed and validated the technology-agnostic heat-engine model, the model can be used to investigate the performance of a heat engine under different operating conditions. For this purpose the parameters listed in Table 4.2 have been defined. Alongside these parameters a non-dimensional heat-source temperature drop has been defined as:

$$\theta = \frac{T_{hi} - T_{ho}}{T_{hi} - T_{ci}}, \qquad (4.27)$$

which can be varied parametrically to investigate the performance of a heat engine under different operating conditions. When $\theta = 0$, there is no reduction in the heat-source temperature, and when $\theta = 1$, the heat source is cooled right down to the heat-sink inlet temperature.

For a range values for θ the heat-engine model was run and the thermodynamic performance of a Carnot and endo-reversible heat engine was determined. The results are shown in Figure 4.4. The plots in (a) and (b) in Figure 4.4 compare the results for a Carnot and endo-reversible heat engine for a fixed heat-sink temperature (i.e. $\alpha \to \infty$) and for a variable heat-sink temperature with $\alpha = 1$. The plots in (c) and (d) show the results for an endo-reversible heat engine with varying values of α.

Table 4.2 Values used to demonstrate the technology-agnostic heat-engine model

Heat-source temperature	T_{hi}	423	K
Heat-source mass flow rate	$\dot{m}_{l_1}$	1.0	kg/s
Heat-source specific-heat capacity	$c_{p,h}$	4,200	J/(kg K)
Heat-sink temperature	T_{ci}	298	K
Operating time per annum	n	8,000	hours
Cost of electricity	C_e	0.05	\$/kWh
Operating and maintenance cost	$C_{o\&m}$	0.01	\$/kWh
Discount rate	r	5	%
Technology lifespan	t_{max}	20	years

Firstly, for a fixed heat sink the maximum power is always obtained when the heat source is fully cooled down to the heat-sink temperature. However, for a variable heat sink, an optimal temperature to which the heat source should be cooled down to is observed. This optimum exists because there is a trade-off between the amount of heat that is added to the thermodynamic cycle, and the amount of heat that must be rejected into the heat sink. This trade-off can explored further by considering Figure 4.5, which shows the temperature profiles within a heat engine, where the heat-exchange processes are co-current and con-current, respectively. In this figure, three cases are considered: (1) T_h = constant and T_c = constant; (2) $T_{ho} = T_{ci}$; and (3) $T_{ho} = T_{co}$. For case (1), constant heat-source and heat-sink temperatures imply a maximum theoretical efficiency but zero work output. On the other hand, case (2) corresponds to a maximum heat transfer between the source and sink. However, since all the thermal energy taken from the heat source is transferred into the sink, this case also corresponds to zero work output. Furthermore, for the co-current heat engine it also observed that this situation violates the constraint $T_c \leq T_h$, and is therefore practically impossible. Finally, it is found that case (3) corresponds to the maximum power point for the heat engine, and it is also observed that this optimal point is independent of whether the heat exchange processes are co-current or con-current.

Referring back to the right-hand plots of Figure 4.4, it is clear that the heat-sink cooling capacity strongly affects the performance of the cycle, with low heat-sink mass-flow rates significantly reducing the potential of the heat engine to produce work. This indicates the importance of considering the available heat sink within the analysis.

Finally, it is clear from Figure 4.4 that for a fully defined heat source and heat sink there is a single optimum for the amount of work that can be produced. Based on this observation, a simple parametric investigation for a range of heat-source temperatures can be completed. For this study the heat-source mass-flow rate, heat-source specific-heat capacity and heat-sink temperature were set to the values given in Table 4.2. The technology-agnostic model, using the endo-reversible efficiency, was then run for a range of heat-source temperatures and values for α and the resulting power output are plotted in Figure 4.6, alongside the optimal values for the non-dimensional

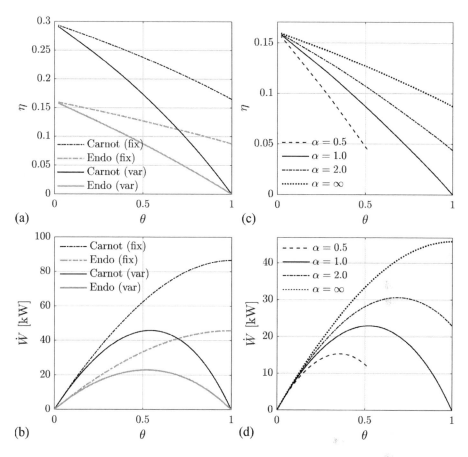

Figure 4.4 Demonstration of the heat-engine technology-agnostic thermodynamic model. Plots (a) and (b) compare the thermodynamic performance of Carnot and endo-reversible heat engines for a fixed (fix) and variable (var) heat-sink temperature, whilst plots (c) and (d) show the performance of an endo-reversible heat engine with a varying heat-sink mass-flow rate

temperature drop θ. Unsurprisingly, the results show that power output increases with an increasing heat-source temperature, and with an increased heat-sink mass-flow rate. On the other hand, the optimal value for θ is only weakly related to the heat-source temperature, but is strongly dependent on the heat-sink mass-flow rate. Ultimately, this figure or a similar figure can be used as a useful tool to predict the performance of an endo-reversible heat engine when both the heat-source temperature and heat-capacity ratio α are known. This figure also reaffirms the importance of considering the available heat sink during heat-engine modelling.

In addition to considering the thermodynamic performance of the heat engine, it is also important to consider the heat-engine economic performance. Referring back to Equation 4.17, it is clear that the annual savings generated from a heat engine are directly proportional to the work generated. Therefore, it immediately follows that the annual savings are maximised at the maximum power point. However, it does not

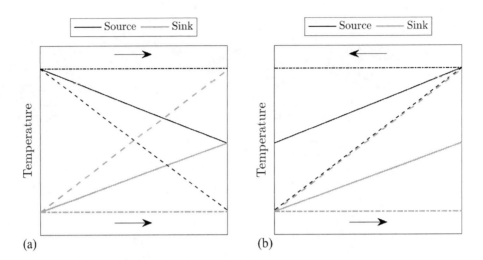

Figure 4.5 Temperature profiles in a co-current (a) and con-current (b) heat engine. The dash-dot lines correspond to case (1), the dashed lines correspond to case (2) and the solid lines correspond to case (3)

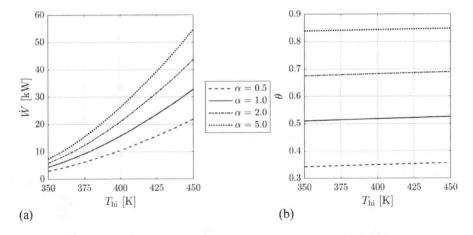

Figure 4.6 The maximum power (a) and optimal non-dimensional temperature drop (b) for an endo-reversible heat engine operating at different heat-source temperatures with different heat-capacity ratios. Results obtained for $\dot{m}_h = 1$ kg/s, $c_{p,h} = 4200$ J/(kg K) and $T_{ci} = 298$ K

immediately follow that this maximum power point corresponds to the minimum SICs or maximum NPV, since this system may in fact require very large heat exchangers. It is therefore necessary to investigate this further.

Within the component cost functions defined in Section 4.1.3, the parameter β is introduced as the ratio of the condenser heat conductance to the evaporator heat conductance (i.e. $\beta = k_c/k_h$), where the heat conductance is also equivalent to the product UA, where U is the overall heat-transfer coefficient and A is the heat-exchanger area. To investigate the effect of β on the economic performance of the heat engine, consider

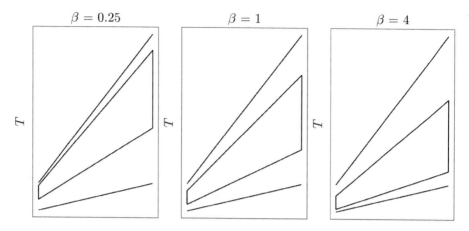

Figure 4.7 The effect of β on the heat-exchanger temperature differences within a heat engine. Results completed for the heat-source and heat-sink conditions defined in Table 4.2 with $\alpha = 5$

Figure 4.7, which shows the working-fluid temperature profile of a heat engine with three different values for β. It should be noted that although the working-fluid temperature profile within each heat engine is different, the work output and thermal efficiency of each heat engine will be the same since the thermodynamic analysis is not dependent on the working-fluid conditions. Hence, the effect of β is purely on the heat-engine economics.

From Figure 4.7 it is observed that as $\beta \to 0$, the temperature difference within the evaporator $\Delta T_h \to 0$. By comparison, as $\beta \to \infty$, the temperature difference within the condenser $\Delta T_c \to 0$. These two conditions imply infinitely large heat exchangers, and therefore also correspond to infinitely large component costs. For the $\beta = 1$ case, it is observed that the temperature differences within the evaporator and condenser are similar and therefore it can be assumed that neither the evaporator nor condenser will be excessively sized. Since the heat-engine power is independent of β, the optimal value for β is determined purely on the economic performance and will correspond to a minimum in the SIC. Figure 4.8 displays this optimum for three different heat-source temperatures and heat-capacity ratios and the results show that for all cases this optimum is observed at $\beta \approx 1$. Therefore, setting $\beta = 1$ seems a reasonable assumption within the technology-agnostic model, and this value is recommended for future investigations.

Having established a suitable value for β and having validated the component cost functions, it is now possible to complete the economic assessment for the demonstration case study defined in Table 4.2. In a similar fashion to the thermodynamic investigation, a parametric investigation of θ was completed for different values of α, and the results are shown in Figure 4.9. The plots in (a) show the resulting thermal efficiency η, power output W and the reciprocal of the specific investment costs SIC^{-1}, all normalised by their maximum values. In these plots, the ecological criterion is also introduced, which is defined as $E = W - T_0\sigma$, where T_0 is the ambient temperature

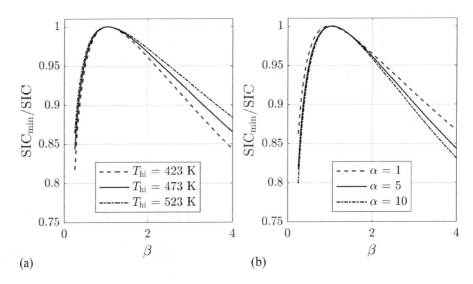

Figure 4.8 The effect of β on the SIC of a heat engine. The results in (a) consider three different heat-source temperatures and the results in (b) consider three different heat-capacity ratios (α)

and σ is the entropy production within the heat engine. The term $T_0\sigma$ corresponds to a power loss, and therefore the optimisation of E represents a compromise between minimising entropy production and maximising the power output. The plots in (c) and (d) show the resulting power output and NPV, also normalised by their maximum values. In these plots, the NPV is determined for three cases, where the net savings per kWh (i.e. $C_n = C_e - C_{o\&m}$) are set to 0.04 \$/kWh, 0.09 \$/kWh and 0.19 \$/kWh, respectively, to investigate how the electricity cost affects the optimum value of θ.

As previously discussed, the thermal efficiency is maximised at $\theta = 0$; however, this case corresponds to $\dot{W} = 0$. The power output $\dot{W}$ is maximised for a specific value of θ, and this optimal value increases as α increases. However, as θ increases, the amount of heat input into the heat engine also increases, which must correspond to larger heat exchangers with higher investment costs. Therefore, when the SIC is considered, it is observed that there is an optimal θ that minimises the SIC, and this optimum is always lower that the θ associated with the maximum power point. Furthermore, since the SIC and the PB for the system are directly related, it immediately follows that the payback is also minimised at this same optimal condition. This therefore indicates that for the minimisation of the SIC, it is better to take less heat from the heat source which could lead to undersized systems. In terms of the ecological criterion, it is observed that this is maximised at a particular value of θ and this value is different again to the values that maximise $\dot{W}$ or minimise the SIC.

Finally, consideration is given due to the NPV, which takes into account the cost of the system in addition to the value of the electricity that is generated over the technology's lifespan. These results show an optimal value of θ that is somewhere in between the value that maximises the power output, but minimises the SIC. Furthermore, the

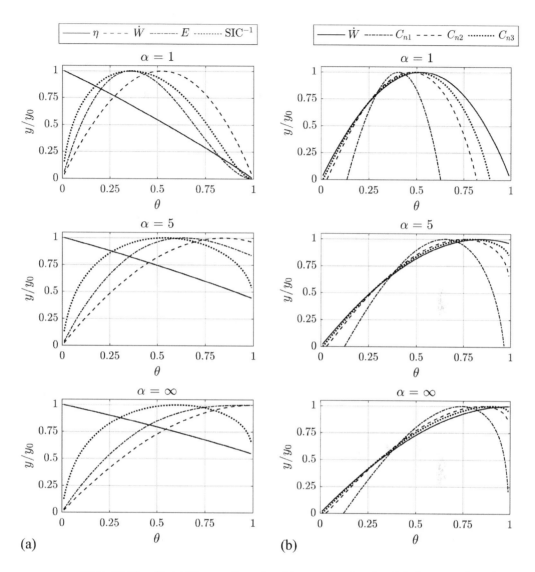

Figure 4.9 The effect of the non-dimensional temperature drop θ and the heat capacity ratio α on the thermodynamic and economic performance of a heat engine. plots in (a) show the thermal efficiency η, power output $\dot{W}$, ecological criterion E and the reciprocal of the specific investment costs SIC^{-1}, and the plots in (b) show the NPV for three values of $C_n = C_e - C_{o\&m}$; $C_{n1} = 0.04$ \$/kWh, $C_{n2} = 0.09$ \$/kWh and $C_{n3} = 0.19$ \$/kWh

optimal θ that maximises the NPV is dependent on the cost of electricity. As the cost of electricity increases, the optimal value for θ tends towards the maximum power point, whilst a reduction in the cost of electricity results in a value of θ that is less than the maximum power point, again indicating a preference towards undersized systems. Ultimately, these results demonstrate that the parameter used to define an 'optimal' system can affect the heat-engine system that is selected.

4.2   Heat-Upgrade Heat Pumps

4.2.1   Thermodynamic Model

In a similar fashion to the heat engine, the optimal performance of a heat-upgrade heat pump operating between a defined heat-source and heat-sink temperature is obtained for a Carnot heat pump, whose coefficient of performance (COP) ϕ is given by:

$$\phi = \frac{T_h}{T_h - T_c}, \tag{4.28}$$

where T_h is temperature of the high-temperature heat sink and T_c is the temperature of low-temperature heat source. As introduced in Section 2.1.6, an endo-reversible heat pump can be modelled using the (Blanchard 1980) correlation:

$$\phi = \left(1 - \frac{T_c}{T_h + \Delta T_k}\right)^{-1}, \tag{4.29}$$

where ΔT_k is an empirical temperature difference, with values of 30 K being recommended, or by using the (Velasco et al. 1997) correlation:

$$\phi = \sqrt{\frac{T_h}{T_h - T_c}}. \tag{4.30}$$

Synonymous with the heat engine, Equations 4.28, 4.29 and 4.30 form the basis of the technology-agnostic heat pump model, and again the analysis is extended by considering varying heat-source and heat-sink temperatures. In this analysis the heat pump is divided into differential heat pumps for which Equation 4.3 again applies (Figure 4.10). The work required by the heat pump is given by:

$$\dot{W} = \dot{m}_c \int_{T_{co}}^{T_{ci}} \frac{c_{p,c}(T_c)}{\phi(T_c) - 1} \, dT_c, \tag{4.31}$$

whilst the heat generated when a heat source is cooled down to particular temperature is given by:

$$\dot{Q}_h = (\dot{m}c_p)_c (T_{ci} - T_{co}) + \dot{W}. \tag{4.32}$$

Again, removing the temperature dependence of the specific-heat capacity and considering a Carnot heat pump, an analytical solution can be obtained:

$$\dot{W} = (\dot{m}c_p)_c \left[(T_{co} - T_{ci}) + \frac{T_{hi}}{\alpha}\left(\left(\frac{T_{ci}}{T_{co}}\right)^\alpha - 1\right)\right]; \tag{4.33}$$

$$\dot{Q}_h = (\dot{m}c_p)_h T_{hi}\left(\left(\frac{T_{ci}}{T_{co}}\right)^\alpha - 1\right); \tag{4.34}$$

$$\phi = \frac{\dot{Q}_h}{\dot{W}} = \alpha \left[1 - \frac{T_{ci} - T_{co}}{T_{hi}\left(\left(\frac{T_{ci}}{T_{co}}\right)^\alpha - 1\right)}\right]^{-1}. \tag{4.35}$$

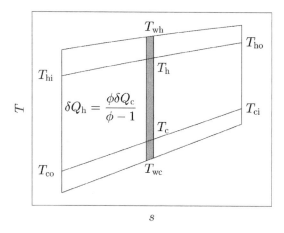

Figure 4.10 Finite-time heat pump operating between a non-isothermal heat source and heat sink represented on a T-s diagram

For an endo-reversible heat pump, using either Equation 4.29 or Equation 4.30, or for a heat source and heat sink with varying specific-heat capacities, the solution to Equations 4.3 and 4.31 can be obtained using a suitable numerical scheme.

Ultimately, these equations help to explain the heat pump performance as the heat-source outlet temperature is reduced. Firstly, at $T_{co} = T_{ci}$ the heat generated by the heat pump, and therefore the work required, are both equal to zero. As T_{co} is reduced, both $\dot{Q}_h$ and $\dot{W}$ will increase, which corresponds to generating more heat, but at the expense of having to put more mechanical work into the system. Therefore, the maximum heat that can be generated by the heat pump corresponds to when the heat source is cooled down to the minimum allowable temperature. However, this operating point will also require the greatest work input. Clearly, when evaluating this from a thermodynamic point of view, an optimal T_{co} doesn't exist. However, when the cost of the heat generated and the cost of the electricity required to drive the heat pump are considered, an optimum will be observed.

As for the heat engine, it is also necessary to predict the temperature differences within the heat pump. Within a heat pump there is a temperature difference between the low-temperature heat source and working fluid, denoted ΔT_c, and a temperature difference between the high-temperature heat sink and the working fluid, denoted ΔT_c, and these can also be seen in Figure 4.10. The COP given by either Equation 4.29 or 4.30 is the endo-reversible COP of the heat pump and therefore already accounts for these temperature differences. Therefore, if the endo-reversible heat pump is thought of as being equivalent to a Carnot heat pump operating between a lower heat-source temperature, denoted T_{wc}, and a higher heat-sink temperature, denoted T_{wc}, it follows that:

$$\phi = \frac{T_{wh}}{T_{wh} - T_{wc}}, \tag{4.36}$$

where ϕ is the COP given by an endo-reversible model. Furthermore, the heat transfer from the heat source to the working fluid is given as:

$$\dot{Q}_c = k_c(T_c - T_{wc}), \tag{4.37}$$

and the heat transfer from the working fluid to the heat sink is calculated as:

$$\dot{Q}_h = k_h(T_{wh} - T_h), \tag{4.38}$$

where k is the heat conductance, which is again equivalent to the product of the overall product of the overall heat-transfer coefficient U and the heat-exchanger area A. Reintroducing $\beta = k_c/k_h$ and solving Equations 4.36, 4.37 and 4.38, the following expressions for T_{wh} and T_{wc} are obtained:

$$T_{wh} = \frac{\frac{\beta \phi T_c}{\phi - 1} + T_h}{\beta + 1}; \tag{4.39}$$

$$T_{wc} = \frac{T_{wh}(\phi - 1)}{\phi}, \tag{4.40}$$

where ϕ is calculated from either 4.29 or 4.30.

Therefore, for a heat pump operating between specified heat-source and heat-sink temperatures, the temperature differences between the heat source and the working fluid, and between the heat sink and the working fluid, can be obtained if the heat-conductance ratio β is known.

4.2.2   Thermodynamic Model Validation

In order to validate the technology-agnostic heat-pump model, the model has been used to predict the performance of commercially available heat pumps, and these predictions can be compared to the performance data quoted by the manufacturer. The results from such a comparison are presented in Figures 4.11 and 4.12. In this analysis, the technology-agnostic heat-pump model was run using three different definitions for the COP, namely, the Carnot COP (Equation 4.28), the COP defined by Blanchard (1980) (Equation 4.29 with $\Delta T_k = 30$ K) and the COP defined by Velasco et al. (1997) (Equation 4.30). Whilst it is clear that it is not suitable to use the Carnot COP, reasonable agreement is observed for the other two definitions. In general, Equation 4.29 results in higher predictions for the COP compared to the manufacturer data, and consequently lower predictions for the required compressor work, whilst Equation 4.30 results in lower predictions for the COP, and a higher compressor work. Therefore, the authors conclude that to obtain a conservative estimate of the heat pump performance, Equation 4.30 should be used, and this is considered to give a reasonable indication of the COP, work input required and the heat generated from the heat pump.

4.2.3   Economic Model

For a heat pump, the annual savings arise from no longer having to generate heat through other means, such as burning natural gas. However, a heat pump requires electricity to operate and therefore the annual savings are given by:

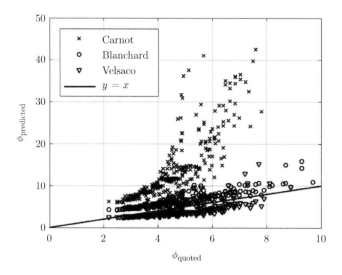

Figure 4.11 Comparison between the manufacturer-quoted COP ϕ_{quoted} and the COP calculated using the technology-agnostic heat-pump model $\phi_{\text{predicted}}$

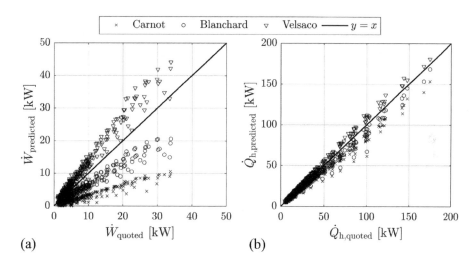

Figure 4.12 Comparison between the manufacturer quoted thermodynamic performance and the performance predicted using the technology-agnostic heat-pump model. (a): power required to drive the heat pump $\dot{W}$; (b): heat produced by the heat pump $\dot{Q}$

$$S = n \left(\dot{Q}_h \left(C_g - C_{o\&m} \right) - \dot{W} C_e \right), \tag{4.41}$$

where $\dot{Q}_h$ and $\dot{W}$ are the heat generated and work required by the heat pump, respectively, and C_g is the cost of natural gas in \$/kWh. The other parameters have the same definition as before.

Table 4.3 Values used to demonstrate the technology-agnostic heat-pump model

Heat-source temperature	T_{ci}	323	K
Heat-source mass-flow rate	$\dot{m}_c$	1.0	kg/s
Heat-source specific-heat capacity	$c_{p,c}$	4,200	J/(kg K)
Heat-sink temperature	T_{hi}	373	K
Ambient temperature	T_0	298	K
Operating time per annum	n	8,000	hours
Cost of natural gas	C_g	0.025	$/kWh
Cost of electricity	C_e	0.05	$/kWh
Operating and maintenance cost	$C_{o\&m}$	0.01	$/kWh
Discount rate	r	5	%
Technology lifespan	t_{max}	20	years

The economic performance of a heat pump is measured using the same performance indicators that were introduced for the heat engine, namely: the SIC, PB, NPV
and LCOE (Equations 4.18-4.21). The only differences when considering a heat pump
is that the SIC is now the cost per kW of heating generated, and $\dot{Q}_h$ is used instead
wherever $\dot{W}$ would be used for a heat engine.

Whilst a heat engine comprises of a pump, expander, evaporator and condenser, a
heat pump consists primarily of a compressor, evaporator and condenser. Therefore,
the only additional cost correlation that is required is for a compressor. The correlation
used for the compressor cost (in $) can be taken from Smith (2005), and takes the form:

$$C_{co} = d_1 \left(\frac{\dot{W}_{co}}{250} \right)^{d_2} , \qquad (4.42)$$

where $d_1 = 9.84 \times 10^4$ and $d_2 = 0.46$.

4.2.4   Application of the Model

To demonstrate the technology-agnostic heat-pump model another case study can be
considered. In this case study, the heat pump is assumed to upgrade the heat from a
low-temperature waste-heat source at 323 K to a high-temperature heat sink at 373 K.
The rest of the assumptions made for the case study are given in Table 4.3. The non-
dimensional heat-source temperature drop is now defined as:

$$\theta = \frac{T_{hi} - T_{ho}}{T_{hi} - T_0} , \qquad (4.43)$$

where T_0 is the ambient or dead-state temperature, and is the minimum temperature
to which the heat source can be cooled down (i.e. $\theta = 1$).

For a range values for θ the technology-agnostic heat-pump model was run
and the performance of a Carnot and endo-reversible heat pump was determined.
Here the endo-reversible model is run using the COP given by Blanchard (1980)

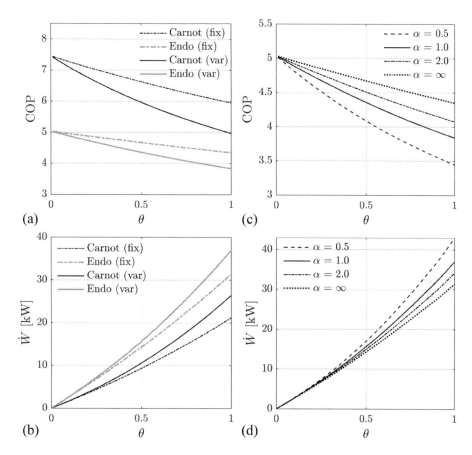

Figure 4.13 Demonstration of the technology-agnostic heat-pump model. (a) and (b) compare the performance of Carnot and endo-reversible heat pumps for a fixed (fix) and variable (var) heat-sink temperature, whilst the (c) and (d) plots show the performance of an endo-reversible heat pump with a varying heat-sink mass-flow rate. For this analysis the Blanchard COP is used with $\Delta T_k = 30$ K

(Equation 4.29). The results are shown in Figure 4.13. Synonymous with the heat-engine example, the plots in (a) and (b) in Figure 4.4 compare the results for the Carnot and the endo-reversible models for a fixed heat-sink temperature (i.e. $\alpha \to \infty$) and for a variable heat-sink temperature with $\alpha = 1$. The plots in (c) and (d) show the results for the endo-reversible model for varying values of α.

Overall it is observed that for all cases an increasing amount of heat into the system corresponds to a reduction in the COP, and also an increase in the required work to drive the heat pump. Since $Q = \phi W$, the heat generated by the heat pump also increases with increasing heat into the system, but this increase is slower than the rate of increase in W due to the reduction in ϕ. As observed for the heat engine, Figure 4.13 also shows that the performance of the heat pump is strongly affected by the mass-flow rate of the heat sink, with a smaller heat-sink mass-flow rate resulting in a lower COP and a greater amount of work required by the system.

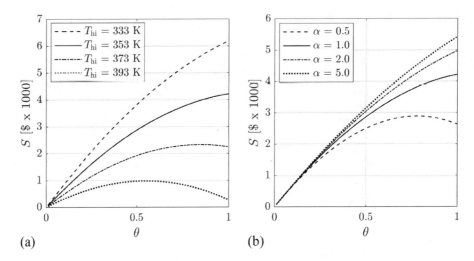

Figure 4.14 The effect of the heat-source outlet temperature on the annual revenue generated by an endo-reversible heat pump, operating with different heat-sink temperatures and heat-sink mass-flow rates. For (a) plot, $\alpha = 1$, and for (b) plot, $T_{hi} = 353$ K. For the analysis, $\Delta T_k = 30$ K is assumed

Unlike the heat-engine example, an optimal performance point is not observed in Figure 4.13. However, when evaluating the performance of the heat pump in terms of the potential revenue generated, it is observed that an optimal operating point exists. This is explored in Figure 4.14, which investigates the effect of the heat-sink conditions on the economic performance of the heat pump. An optimum is observed because an increasing heat input into the heat pump translates into a greater heat output and a reduction in natural gas consumption. However, this increasing input also translates into a greater work requirement, and therefore a greater electricity consumption. Clearly, there is trade-off between maximising natural gas savings whilst minimising electricity consumption, and this trade-off leads to an optimal value for θ. It is also interesting from Figure 4.14 that maximum revenue is generated when a large mass-flow rate is upgraded by a small temperature.

4.3   Absorption Chillers

4.3.1   Thermodynamic Model

Typically, to model an absorption chiller the whole system is thought of as the combination of a heat engine and a heat pump, and a third temperature level is introduced (Yan & Chen 1989; Abrahamsson & Jernqvist 1993; Chen 1995). The heat engine operates between the heat source and ambient heat sink, and the work generated by this heat engine drives the heat pump which transfers heat from the cold sink to the ambient heat sink. Therefore, the COP of an absorption chiller ψ is the product of the heat-engine thermal efficiency η and the heat-pump COP. Finally, it should also be noted that the COP of a refrigeration heat pump and a heat-upgrade heat pump are related by $\phi_{\text{heating}} = \phi_{\text{cooling}} + 1$, hence:

$$\psi = \frac{\dot{Q}_c}{\dot{Q}_h} = \frac{\dot{W}}{\dot{Q}_h}\frac{\dot{Q}_c}{\dot{W}} = \eta(\phi - 1), \tag{4.44}$$

where $\dot{Q}_h$ is the heat supplied to absorption chiller and $\dot{Q}_c$ is the amount of cooling provided by the absorption-chiller system. For an ideal Carnot absorption chiller operating between isothermal heat-source and heat-sink temperatures, the COP is given by:

$$\psi = \left(1 - \frac{T_0}{T_h}\right)\left(\frac{T_c}{T_0 - T_c}\right), \tag{4.45}$$

where T_h, T_c and T_0 are the temperatures of the high-temperature heat source, low-temperature heat sink and ambient heat sink, respectively. Similarly, the COP for an endo-reversible absorption chiller operating between an isothermal heat source and isothermal heat sinks is given by:

$$\psi = \left(1 - \sqrt{\frac{T_0}{T_h}}\right)\left(\left(1 - \frac{T_c}{T_0 + \Delta T_k}\right)^{-1} - 1\right), \tag{4.46}$$

when using the Blanchard (1980) correlation to model the endo-reversible heat pump, and

$$\psi = \left(1 - \sqrt{\frac{T_0}{T_h}}\right)\left(\sqrt{\frac{T_0}{T_0 - T_c}} - 1\right), \tag{4.47}$$

when using the Velasco et al. (1997) correlation to model the endo-reversible heat pump.

The additional complexity of introducing the third temperature level means that it is not possible to generate analytical expressions for ψ for a non-isothermal heat source and non-isothermal heat sinks. However, by decoupling the system into a separate heat pump and heat engine it is simpler to evaluate the performance of the system. In Figure 4.15, a schematic of the decoupled technology-agnostic absorption chiller is shown for the non-isothermal condition. For a known heat-source temperature drop (i.e. $T_{hi} - T_{ho}$) the power output from the heat engine can be calculated using the heat-engine technology-agnostic model. Furthermore, for a defined heat-sink temperature drop (i.e. $T_{ci} - T_{co}$) the power required to drive the heat pump can be calculated using the heat-pump technology-agnostic model. It is then a matter of varying the heat-sink temperature drop within an iterative loop until the power generated by the heat engine and the power required by the heat pump match, at which point the actual heat-sink outlet temperature for the imposed heat-source temperature drop is known.

4.3.2 Thermodynamic Model Validation

As for the heat engine and heat pump models, it is also possible to validate the technology-agnostic absorption-chiller model by comparing predictions to performance data taken from manufacturer data sheets. After collecting data from various

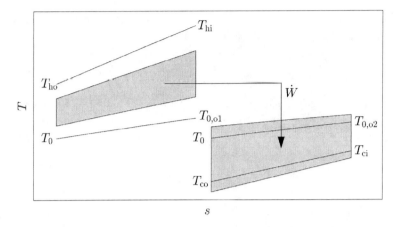

Figure 4.15 Finite-time absorption chiller operating between a non-isothermal heat source and heat sink represented on a T–s diagram

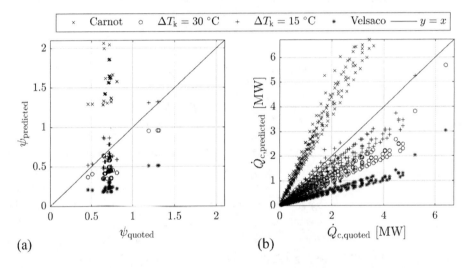

Figure 4.16 Comparison manufacturer-quoted performance data and performance predictions obtained using the technology-agnostic absorption-chiller model. (a): COP ψ; (b): cooling capacity $\dot{Q}_c$

manufacturers, the technology-agnostic model was initially run three times. The first case corresponded to a Carnot absorption chiller (Equation 4.45), the second corresponded to using the endo-reversible model with Equation 4.46, where $\Delta T_k = 30\ °C$, and the third corresponded to using the endo-reversible model with Equation 4.47. The results from this comparison are shown in Figure 4.16.

From the results in Figure 4.16 it is clear that using the Carnot COP in the technology-agnostic model predicts significantly higher COPs and cooling capacities than the values quoted by the manufacturers. On the other hand, both endo-reversible models predict much lower COPs and cooling capacities than the values quoted by the manufacturers. Overall, this indicates that none of the three models, in their current state, provide suitably accurate predictions for the heat pump performance, and cannot be relied upon to give accurate predictions of expected performance. However, it is noted that the when using the endo-reversible model based on Equation 4.46, there exists an empirical factor, ΔT_k, which provides an additional degree of control over the technology-agnostic model. Indeed, when this parameter is reduced to $\Delta T_k = 15\,°C$, it is observed that a more reasonable agreement between the technology-agnostic model and the manufacturer-quoted data is obtained, as observed in Figure 4.16. This suggests that the value originally quoted by Blanchard (1980) may not be suitable for modelling an absorption-chiller system, which is not surprising as the original correlation was never intended for application to these systems. Instead, it appears that $\Delta T_k = 15\,°C$ may be more suitable, although further validation of this is required.

4.3.3   Economic Model

For an absorption chiller the energy savings can be related to the equivalent work that would be required to provide the same refrigeration effect using a mechanical vapour-compression heat pump. This is equivalent to the work that is generated by a heat engine operating between the defined heat source and ambient heat sink, and this is known from the decoupled absorption-chiller technology-agnostic model. The annual savings are therefore given by:

$$S = n(\dot{W}C_e - \dot{Q}_c C_{o\&m}),\qquad(4.48)$$

where $\dot{W}$ is the required work and $\dot{Q}_c$ is the cooling generated by the absorption chiller. The remaining parameters have the same definitions as before.

The economic performance of an absorption chiller can also be measured by the SIC, PB, NPV and LCOE (Equations 4.18-4.21), noting that the SIC is now the cost per kW of cooling generated, and $\dot{Q}_c$ is used instead wherever $\dot{W}$ would be used for a heat engine. The main components that make up an absorption chiller are four exchangers, and therefore the heat-exchanger cost correlations described for the heat engine can also be applied here.

4.3.4   Application of the Model

Having now set up and validated the technology-agnostic absorption-chiller model, the model can be used to investigate the performance of an absorption chiller under different operating conditions. For this purpose the parameters listed in Table 4.4 have been defined. Compared to the heat-engine and heat-pump model, an additional heat sink is introduced, to which both the heat engine and heat pump reject heat, and this is defined by T_0, $c_{p,0}$ and $\dot{m}_0$. The two heat-rejection streams that interact with the heat engine and heat pump, respectively, are assumed to be under the same conditions, but are assumed to be independent of one another. For the analysis the non-dimensional heat-source temperature drop has been defined as:

$$\theta = \frac{T_{hi} - T_{ho}}{T_{hi} - T_0},\qquad(4.49)$$

Table 4.4 Values used to demonstrate the technology-agnostic absorption-chiller model

Heat-source temperature	T_{hi}	423	K
Heat-source mass flow rate	$\dot{m}_h$	1.0	kg/s
Heat-source specific-heat capacity	$c_{p,h}$	4,200	J/(kg K)
Heat-sink temperature	T_{ci}	278	K
Heat-sink mass-flow rate	$\dot{m}_c$	5.0	kg/s
Heat-sink specific-heat capacity	$c_{p,c}$	4,200	J/(kg K)
Ambient-sink temperature	T_0	298	K
Ambient-sink mass-flow rate	$\dot{m}_0$	5.0	kg/s
Ambient-sink specific-heat capacity	$c_{p,0}$	4,200	J/(kg K)
Operating time per annum	n	8,000	hours
Cost of electricity	C_e	0.05	\$/kWh
Operating and maintenance cost	$C_{o\&m}$	0.01	\$/kWh
Discount rate	r	5	%
Technology lifespan	t_{max}	20	years

which can be varied parametrically to investigate the performance of an absorption chiller under different operating conditions.

For a range of values for θ the technology-agnostic absorption-chiller model was run considering both a Carnot system and an endo-reversible system. For the heat-pump cycle the Blanchard (1980) correlation was used with $\Delta T_k = 30$ K. The results are shown in Figure 4.17, and again, following the previous format, the plots in (a) and (b) in Figure 4.17 compare the results for fixed and variable heat sinks, whilst plots in (c) and (d) show the results for the endo-reversible model for varying values of α. It should be noted that since there are now multiple heat sinks, three cases are considered in (a) and (b) plots: (1) 'f' – both the cold and ambient heat sinks are fixed; (2) 'vf' – the cold and ambient heat sinks are variable and fixed, respectively; (3) 'vv' – the cold and ambient heat sinks are both variable. In the case of variable heat sinks, $\alpha = 5$ is assumed.

From Figure 4.17 it is observed that for an increasing amount of heat input into the system, the COP ψ reduces, and also that the variable heat sink and endo-reversible systems result in significantly lower COPs than the fixed heat-sink Carnot system. Furthermore, when considering an isothermal ambient heat sink, but a variable cold heat sink, an optimal performance point is not observed (i.e. the maximum cooling capacity is generated when the heat source is cooled down to the ambient temperature). However, when considering a variable ambient heat sink, it is observed that there is an optimal value of θ at which a maximum cooling capacity is generated. It follows that this point will also correspond to a maximum in energy savings. Furthermore, replacing an existing refrigeration unit with a heat to cooling system reduces electricity consumption. Therefore, since energy savings are related to this reduction in electricity consumption, it also follows that the optimal point corresponds to the point at which the heat engine that drives the heat pump generates the maximum power. In other words, it can be shown that the economic benefit in an absorption system

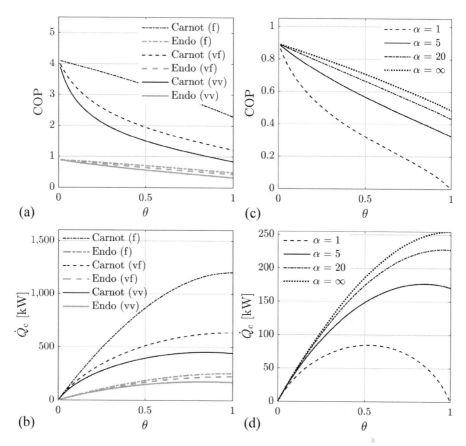

Figure 4.17 Demonstration of the technology-agnostic absorption-chiller model. The plots in (a) and (b) compare the performance of a Carnot system and an endo-reversible system for fixed and variable heat-sink temperatures, whilst the plots in (c) and (d) show the performance of an endo-reversible system operating with varying heat-sink mass flow rates. For the heat pump endo-reversible model, $\Delta T_k = 30$ K was assumed

is essentially the same as a heat engine operating between the heat source and the ambient heat sink. The only difference is whether the work generated is used to provide refrigeration, or used as electricity. The decision on which technology is most suitable should be based on whether there is a greater need for electricity or cooling on-site, and also on the required investment cost for the heat engine or absorption-chiller system.

To further widen the analysis, it is also interesting to use the technology-agnostic model to investigate how the performance of the system is likely to change as the heat source, heat sink and ambient temperatures vary. To this end, three parametric studies have been completed and the results obtained are reported in Figures 4.18–4.20.

In terms of the effect of the heat-source temperature, it is observed that an increase in temperature results in the chiller being able to generate more cooling power. This is not particularly surprising, since a higher heat-source temperature corresponds to

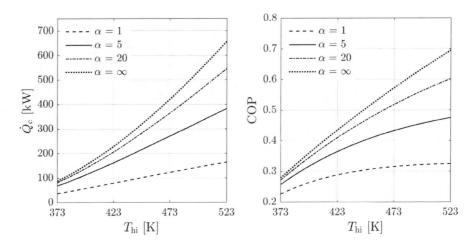

Figure 4.18 Effect of heat-source temperature on the performance of an absorption chiller assuming $T_{ci} = 273$ K, $T_0 = 298$ K and $\alpha = (\dot{m}c_p)_c/(\dot{m}c_p)_h = (\dot{m}c_p)_a/(\dot{m}c_p)_c$

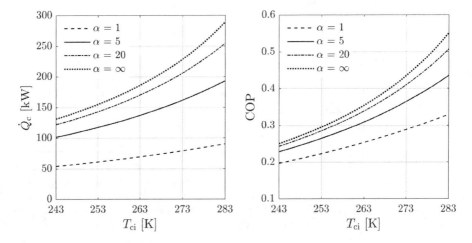

Figure 4.19 Effect of heat-sink temperature on the performance of an absorption chiller assuming $T_{hi} = 423$ K, $T_0 = 298$ K and $\alpha = (\dot{m}c_p)_c/(\dot{m}c_p)_h = (\dot{m}c_p)_a/(\dot{m}c_p)_c$

more heat being available, and one would expect this to lead to a larger cooling effect. However, it is also observed that as heat-source temperature increases, the COP of the chiller also increases. In essence this means that not only there is more heat available, but also the system is able to more effectively convert this heat into cooling. Thus, it can be concluded that a heat-driven chiller becomes more effective as the temperature of the heat source increases. Although the technology-agnostic model may not be completely representative of a real system, this can be understood by considering that in effect a higher heat-source temperature leads to a more efficient heat engine, which in turn leads to more power being generated, which then drives the refrigeration system. Finally, we see that increasing the size of the ambient and cold heat sinks, relative to the heat source, leads to a larger cooling capacity, and also a higher COP.

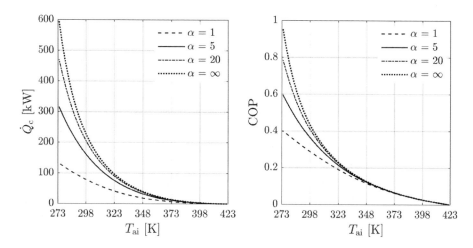

Figure 4.20 Effect of ambient temperature on the performance of an absorption chiller assuming $T_{hi} = 423$ K, $T_{ci} = 273$ K and $\alpha = (\dot{m}c_p)_c/(\dot{m}c_p)_h = (\dot{m}c_p)_a/(\dot{m}c_p)_c$

Moving on to consider the effect of the heat-sink temperature, it is observed that an increase in the heat-sink temperature corresponds to an increase in both the cooling capacity and the COP of the system. This may seem slightly counter-intuitive; in the previous paragraph we have seen that higher heat-source temperatures lead to better performance and thus one might be forgiven for suggesting that performance increases as the temperature difference between the heat source and heat sink increases. However, the behaviour observed can be explained by again considering that the refrigeration part of the agnostic model is modelled as a vapour-compression refrigeration system, and we have already seen that the COP of a vapour-compression refrigeration system increases as the temperature difference between the two reservoirs reduces. Thus, it is better to revise that statement, saying instead that the technology-agnostic model suggests that the performance of a heat-driven chiller increases as the temperature difference between the heat source and ambient conditions increases, and as the temperature difference between the cold sink and ambient conditions reduces.

This observation is further confirmed by considering the results from the parametric study concerning the effect of the ambient temperature (Figure 4.20). Again, we see a significant increase in the COP for the heat-driven chiller system, as the ambient temperature approaches the temperature of the cold reservoir. At this point, both the efficiency of the heat engine and the COP of the heat pump are maximised. As a final observation, for both Figures 4.19 and 4.20, we again see that increasing the relative size of the ambient and cold sinks improves the performance of the system.

4.4   Summary

Having identified a potential heat source, which could be utilised and converted into useful forms of energy such as heating, cooling or electricity, one of the first steps in identifying the feasibility of the system is to assess the amount of energy that could be feasibly produced. Often at this stage it is impractical to conduct a detailed simulation of different technologies, and instead simple methods to determine this information are required. To this end, this chapter has presented a number of technology-agnostic models that can be used to conduct a first-stage assessment of different heat-utilisation technologies. These models are capable of estimating the thermodynamic performance, be that the thermal efficiency and power output from a heat engine, the COP and heating capacity from a heat pump or the COP and cooling capacity from a heat-driven chiller, from only the known heat-source conditions, heat-sink conditions and ambient conditions. Moreover, when coupled to additional plant information, such as current energy costs and operating hours, these models can also be used to assess potential economic performance.

Within this chapter, these models have been described and validated against more sophisticated models, and performance data where available. However, it is worth emphasising that in developing these models it is necessary to make a number of simplifying assumptions, which means that the models presented may not be completely representative of real systems or their performances. Nonetheless, these models are considered to be a useful tool to identify the most promising heat sources, alongside the most suitable heat-utilisation technology, without requiring those involved in making such decisions to develop more sophisticated models. Having then identified the most feasible options, it is of course impossible to avoid conducting a more detailed evaluation of the heat-utilisation technology selected. Thus, the detailed analysis of these systems will be covered in the following two chapters.

5   Rankine Cycles

From the review of waste-heat recovery technologies completed in Chapter 3 it is apparent that heat to power systems based on the Rankine cycle are promising candidates for waste-heat recovery. Furthermore, the technology-agnostic heat-engine model developed in Chapter 4 can provide a reasonable first-stage assessment of the expected thermoeconomic performance of a heat engine, which in turn can help site managers and engineers to identify potential waste-heat streams that can be utilised. Having identified a potential waste-heat stream, the next step is to investigate in more detail the implementation of a Rankine cycle-based waste-heat recovery system. The aim of this chapter is therefore to provide the reader with a thorough overview of the application of Rankine cycles, operating with both water and organic fluids, for the conversion of waste heat into power. This includes details such as thermodynamic modelling, working-fluid selection, system optimisation and finally component selection.

5.1   Thermodynamic Modelling of the Rankine Cycle

A detailed investigation into the implementation of a Rankine cycle for waste-heat recovery is based on a thermodynamic analysis of the cycle. The purpose of the thermodynamic cycle analysis is to determine the thermodynamic properties of the working fluid within the cycle and to determine the thermodynamic performance of the cycle when operating with a defined heat source and heat sink. Essentially, this is completed by applying an energy balance to each component within the cycle. The aim of this section is to guide the reader through this process.

5.1.1   Thermodynamic Properties

A Rankine cycle is a thermodynamic cycle in which a working fluid undergoes a series of thermodynamic processes to convert heat into work. Within each process the thermodynamic properties of the working fluid change and therefore the most important requirement for conducting a thermodynamic cycle analysis is having a suitable equation of state. This equation of state must accurately describe the state properties (i.e. temperature T, pressure P, enthalpy h, entropy s, density ρ and vapour quality q) of the working fluid over the entire operating region of the cycle. There are various equations of state that could be employed for this purpose, but by far the most commonly applied method is the use of NIST REFPROP (Lemmon, Huber & McLinden 2013), which is a program capable of accurately predicting the fluid properties for a large range of working fluids. CoolProp is open-source alternative that is also commonly used (Bell et al. 2014). These two programs employ a range of different equations of state that contain a large number of parameters that enable excellent correlations to experimental data. Other methods include employing cubic equations of state, such as the Peng-Robinson (Peng & Robinson 1976) and Redlich-Kwong-Soave (Soave 1972), whilst, more recently, advanced fluid models based on statistical associating fluid theory (SAFT) (Chapman et al. 1990) have been used to study organic Rankine cycle systems (Lampe et al. 2015; Oyewunmi et al. 2016). These latter models have the advantage of being able to model fluids for which experimental data may not be available.

To avoid a loss of generality, the thermodynamic analysis presented in this chapter will be formulated for a generic equation of state. Within a thermodynamic analysis all of the thermodynamic properties of a fluid at a particular state can be determined if two properties are known. Expressed mathematically, this can be written as:

$$\mathbf{y} = \mathrm{EoS}(x_1, x_2, \mathrm{fluid}), \tag{5.1}$$

where $\mathbf{y}$ is a vector of the desired thermodynamic properties, x_1 and x_2 are the two known properties and 'fluid' is a group of parameters which define the working fluid.

5.1.2   The Simple Rankine Cycle

We consider first a simple Rankine cycle in which the working fluid is expanded from a saturated-vapour or superheated state. Furthermore, the cycle is subcritical (i.e. the operating pressures are below the critical pressure of the working fluid), and the system does not include a recuperator. Therefore, the system is comprised of a pump, evaporator, expander and condenser. This cycle is shown in Figure 5.1, in which the notation used to describe the cycle is also shown.

Alongside the assumptions that define the cycle architecture, a number of additional assumptions are made, and can be summarised as follows:

- Pressure losses within the cycle are neglected;
- Heat losses to the surrounding are neglected;
- The cycle is assumed to operate under steady conditions;
- The working fluid at the pump inlet is assumed to be a saturated liquid;
- No phase change of the heat source or heat sink is permitted. Therefore, the heat source and heat sink are fully defined in terms of their:

    - inlet temperatures, T_{hi} and T_{ci} in K;
    - mass-flow rates, $\dot{m}_{\mathrm{h}}$ and $\dot{m}_{\mathrm{c}}$ in kg/s;
    - specific-heat capacities, $c_{p,\mathrm{h}}$ and $c_{p,\mathrm{c}}$ in J/(kg K);
    - fluid compositions.

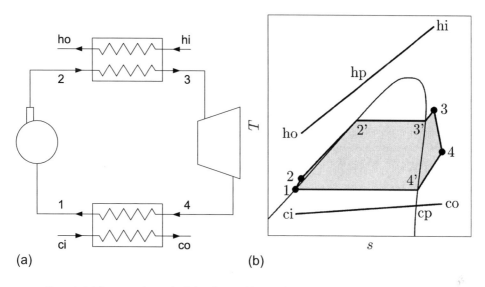

(a)   (b)

Figure 5.1 The operating principle of a Rankine cycle described in terms of the system components (a) and the thermodynamic cycle represented on a temperature-entropy diagram (b)

In addition to defining the working fluid of choice, a non-recuperated, subcritical Rankine cycle can be fully defined by three thermodynamic parameters. Two are needed to define the evaporation and condensation pressures, and a third determines the conditions of the working fluid at the inlet to the expander. There are various ways of specifying these, but the method described here is based on the following three parameters:

- the condensation temperature, T_1 in K;
- the reduced pressure, $P_r = P_2/P_{cr}$, where P_2 is the evaporation pressure and P_{cr} is the critical pressure;
- the amount of superheat, ΔT_{sh} in K.

In addition to these three parameters, four parameters are also required to describe the performance of the each system component, namely:

- the pump isentropic efficiency, η_p;
- the expander isentropic efficiency, η_e;
- the evaporator pinch point, PP_h in K;
- the condenser pinch point, PP_c in K.

Our analysis starts at the pump inlet. Based on our assumption that the working fluid is compressed from a saturated liquid, the working-fluid properties at the pump inlet (enthalpy h_1, entropy s_1 and pressure P_1) are given by the defined condensation temperature T_1 and by setting the vapour quality to zero:

$$[P_1, h_1, s_1] = \text{EoS}(T_1, q_1 = 0, \text{fluid}).\tag{5.2}$$

The pump outlet pressure P_2 is defined by the evaporator reduced pressure:

$$P_2 = P_r \times P_{cr}, \tag{5.3}$$

from which the enthalpy of the working fluid following an isentropic compression (i.e. $s_2 = s_1$) can be calculated:

$$h_{2s} = \text{EoS}(P_2, s_1, \text{fluid}). \tag{5.4}$$

In reality the pump compression is not isentropic, and this is captured by the pump isentropic efficiency, which is defined as the ratio of the pump isentropic work to the actual work:

$$\eta_p = \frac{h_{2s} - h_1}{h_2 - h_1}, \tag{5.5}$$

from which the actual enthalpy h_2 and other thermodynamic properties follow:

$$h_2 = h_1 + \frac{h_{2s} - h_1}{\eta_p}; \tag{5.6}$$

$$[T_2, s_2] = \text{EoS}(P_2, h_2, \text{fluid}). \tag{5.7}$$

In the evaporator the heat source is used to heat up the working fluid to the desired expander inlet conditions. In the case of a superheated cycle this corresponds to the heat-source preheating, fully evaporating and then superheating the working fluid. This heat addition to the cycle occurs under isobaric conditions and therefore $P_2 = P_3$. The working-fluid properties at the start of evaporation ($2'$), end of evaporation ($3'$) and the evaporated outlet are then determined as follows:

$$[T_{2'}, h_{2'}, s_{2'}] = \text{EoS}(P_2, q_{2'} = 0, \text{fluid}); \tag{5.8}$$

$$[T_{3'}, h_{3'}, s_{3'}] = \text{EoS}(P_3, q_{3'} = 1, \text{fluid}); \tag{5.9}$$

$$T_3 = T_{3'} + \Delta T_{sh}; \tag{5.10}$$

$$[h_3, s_3] = \text{EoS}(T_3, P_3, \text{fluid}). \tag{5.11}$$

In the expander the working fluid is expanded back down to the condensation pressure (i.e. $P_4 = P_1$). The enthalpy of the working fluid following an isentropic expansion (i.e. $s_4 = s_3$) can then be calculated as:

$$h_{4s} = \text{EoS}(P_1, s_3, \text{fluid}). \tag{5.12}$$

As for the pump, a real expansion process will not be isentropic, and this behaviour can again be captured by the expander isentropic efficiency, defined as:

$$\eta_e = \frac{h_3 - h_3}{h_3 - h_{4s}}, \tag{5.13}$$

and is the ratio of the actual expander work to the isentropic work. The enthalpy at the expander outlet and remaining thermodynamic properties then follow:

$$h_4 = h_3 - \eta_e(h_3 - h_{4s}); \tag{5.14}$$

$$[T_4, s_4] = \text{EoS}(P_1, h_4, \text{fluid}). \tag{5.15}$$

Finally, the working fluid passes through the condenser where it rejects heat to the heat sink. During this isobaric heat-rejection process the working fluid is precooled to the saturation temperature and it fully backs down to the saturated-liquid pump inlet condition. The working-fluid properties at the start of condensation (4') are given as follows:

$$[T_{4'}, h_{4'}, s_{4'}] = \text{EoS}(P_1, q_{4'} = 1, \text{fluid}). \tag{5.16}$$

Having determined the properties of the working fluid at all the state points within the cycle, the thermodynamic performance can be evaluated. Applying the first law of thermodynamics to the pump, the pump specific work is then:

$$w_p = \frac{\dot{W}_p}{\dot{m}_{wf}} = h_2 - h_1, \tag{5.17}$$

where $\dot{W}_p$ is the power required by the pump and $\dot{m}_{wf}$ is the working-fluid mass-flow rate. The expander-specific work is:

$$w_e = \frac{\dot{W}_e}{\dot{m}_{wf}} = h_3 - h_4, \tag{5.18}$$

where $\dot{W}_e$ is the power generated by the expander. Similarly, for the evaporator, the heat transferred into the cycle per unit mass is:

$$q_h = \frac{\dot{Q}_h}{\dot{m}_{wf}} = h_3 - h_2, \tag{5.19}$$

and the heat rejected from the cycle per unit mass is:

$$q_c = \frac{\dot{Q}_c}{\dot{m}_{wf}} = h_4 - h_1, \tag{5.20}$$

where $\dot{Q}_h$ and $\dot{Q}_c$ are the total heat-transfer rates in the evaporator and condenser, respectively.

The net specific work and thermal efficiency of the cycle then follow as:

$$w_n = w_e - w_p; \tag{5.21}$$

$$\eta_{th} = \frac{w_e - w_p}{q_h}. \tag{5.22}$$

To determine the total work produced by the cycle it is necessary to determine the working-fluid mass-flow rate, which in turn is dependent on the heat-source conditions. Applying an energy balance to the evaporator one obtains:

$$\dot{m}_{wf}(h_3 - h_2) = \dot{m}_h c_{p,h}(T_{hi} - T_{ho}), \tag{5.23}$$

where T_{ho} is the heat-source outlet temperature. Unfortunately, this equation contains two unknowns, namely $\dot{m}_{wf}$ and T_{ho}, and therefore cannot be solved for $\dot{m}_{wf}$. However, the evaporator pinch point is defined as a model input, and this determines the heat-source temperature at the start of the evaporation:

$$T_{hp} = T_{2'} + PP_h. \tag{5.24}$$

Therefore, applying an energy balance to the evaporation and superheating processes only one obtains:

$$\dot{m}_{\text{wf}}(h_3 - h_{2'}) = \dot{m}_{\text{h}}c_{p,\text{h}}(T_{\text{hi}} - T_{\text{hp}}), \tag{5.25}$$

in which $\dot{m}_{\text{wf}}$ is the only unknown. Having obtained $\dot{m}_{\text{wf}}$, T_{ho} then follows from the solution to Equation 5.23. Having determined the heat-source outlet temperature, it is necessary to check that the temperature difference at the evaporator inlet does not violate the pinch-point constraint:

$$T_{\text{ho}} - T_2 \geq PP_{\text{h}}, \tag{5.26}$$

and similarly, the temperature difference at the evaporator outlet must not violate the pinch-point constraint:

$$T_{\text{hi}} - T_3 \geq PP_{\text{h}}. \tag{5.27}$$

The heat-sink temperature at the condenser pinch point is determined from an energy balance:

$$T_{\text{cp}} = T_{\text{ci}} + \frac{\dot{m}_{\text{o}}(h_{4'} - h_1)}{\dot{m}_{\text{c}}c_{p,\text{c}}}, \tag{5.28}$$

and the heat-sink temperature leaving the condenser is similarly determined:

$$T_{\text{co}} = T_{\text{ci}} + \frac{\dot{m}_{\text{o}}(h_4 - h_1)}{\dot{m}_{\text{c}}c_{p,\text{c}}}. \tag{5.29}$$

The temperature differences within the condenser must then all be greater than specific condenser pinch point:

$$T_1 - T_{\text{ci}} \geq PP_{\text{c}}; \tag{5.30}$$

$$T_{4'} - T_{\text{cp}} \geq PP_{\text{c}}; \tag{5.31}$$

$$T_4 - T_{\text{co}} \geq PP_{\text{c}}. \tag{5.32}$$

Finally, having completed the cycle analysis, the net power produced by the cycle is given by:

$$\dot{W}_{\text{n}} = \dot{m}_{\text{wf}}w_{\text{n}}. \tag{5.33}$$

5.1.3   The Recuperated Rankine Cycle

Referring back to Figure 5.1 it is apparent that after expansion the working fluid is superheated. In this instance it is possible to improve the thermal efficiency of the system by adding a recuperator into the system. For clarity the schematic of the recuperated cycle shown in Chapter 3 is shown again here in Figure 5.2.

As observed in Figure 5.2, the purpose of the recuperator is to use the superheated vapour that is leaving the expander to preheat the subcooled liquid working that is leaving the pump. An energy balance applied to the recuperator results in:

$$h_{2\text{r}} - h_2 = h_4 - h_{4\text{r}}, \tag{5.34}$$

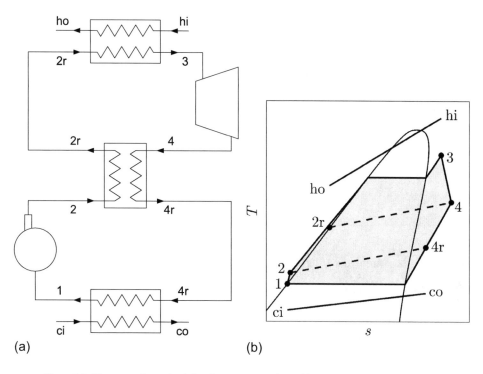

Figure 5.2 The operating principle of a recuperated Rankine cycle described in terms of the system components (a) and the thermodynamic cycle represented on a temperature–entropy diagram (b)

where h_{2r} and h_{4r} are the enthalpies of the working fluid at the two outlets of the recuperator. These two enthalpies are linked by the recuperator effectiveness, which can be defined as a model input:

$$\varepsilon_r = \frac{h_{2r} - h_2}{\Delta h_{max}} = \frac{h_4 - h_{4r}}{\Delta h_{max}}, \tag{5.35}$$

where Δh_{max} is the maximum possible change in enthalpy of the working fluid within the recuperator. This will occur when either the subcooled liquid leaving the pump is heated up to T_4 or when the superheated vapour leaving the expander is cooled down to T_2, hence:

$$h_{2r,max} = \text{EoS}(P_2, T_4, \text{fluid}); \tag{5.36}$$

$$h_{4r,min} = \text{EoS}(P_1, T_2, \text{fluid}); \tag{5.37}$$

$$\Delta h_2 = h_{2r,max} - h_2; \tag{5.38}$$

$$\Delta h_4 = h_4 - h_{4r,min}. \tag{5.39}$$

The parameter Δh_{max} is then given by the smallest of these two values (i.e. Δh_2 or Δh_4), since the stream that requires the smallest change in enthalpy to reach the maximum or minimum temperature will reach that temperature first, hence:

$$\Delta h_{\max} = \min\{\Delta h_2, \ \Delta h_4\}. \tag{5.40}$$

When considering a recuperator within a Rankine cycle the basic calculation procedure can be summarised as follows:

- Conduct the same cycle analysis as described for the simple Rankine cycle to determine state points '1' to '4' (Equations 5.2-5.18);
- Complete the recuperator analysis described in this section for a specified recuperator effectiveness to determine state points '2r' and '4r';
- Complete the remainder of the cycle analysis described for the simple Rankine cycle (Equations 5.19-5.33), replacing state point '2' with '2r' and state point '4' with '4r'.

5.1.4 A Rankine Cycle Operating with a Working-Fluid Mixture

In a Rankine cycle operating with a mixture, the system construction and thermodynamic analysis remains the same. Therefore, the model is easily adapted for modelling a Rankine cycle operating with a mixture of two working fluids. This is completed by specifying two working fluids instead of a single working fluid, and introducing another parameter x, which describes the composition of the mixture. For example, $x = 0.25$ would correspond to a mixture composed from 25% of fluid A and 75% of fluid B. Following the general notation introduced for the equation of state, this can be expressed mathematically as:

$$\mathbf{y} = \mathrm{EoS}(x_1, x_2, \mathrm{fluid_A}, \mathrm{fluid_B}, [x \ \ 1-x]). \tag{5.41}$$

The main requirement for modelling Rankine cycles operating with a fluid mixture is to have a thermodynamic fluid model that is capable of determining the thermodynamic properties of a mixture of two or more fluids. Mixture properties for certain mixtures are readily available in REFPROP and CoolProp, whilst cubic equations of state and models based on SAFT can also be extended to mixtures with the implementation of suitable mixing rules. This is outside the scope of this book, but more information can be found in Poling, Prausnitz and O'Connell (2001).

5.1.5 Trilateral or Partially Evaporated Rankine Cycles

In a trilateral or partially evaporated cycle, expansion occurs from either a saturated liquid (trilateral) or a two-phase state. In this instance, the simple ORC model can be adapted by removing the superheat parameter ΔT_{sh} and replacing this with a parameter that describes the vapour quality q_3 of the fluid at the expander inlet. In this case, $q_3 = 0$ corresponds to a trilateral cycle, whilst $0 < q_3 \leq 1$ corresponds to a partially evaporated cycle. Therefore, the expander inlet conditions are determined from:

$$[h_3, s_3] = \mathrm{EoS}(P_3, q_3, \mathrm{fluid}). \tag{5.42}$$

5.1.6   The Supercritical Rankine Cycle

In the supercritical cycle, expansion occurs from a pressure that is greater than the fluid critical pressure (i.e. $P_2 > P_{cr}$). The cycle analysis can still be modelled by defining the condensation temperature T_1 and by the reduced pressure P_r, noting that for a supercritical cycle the high-pressure level is no longer referred to as the evaporation pressure as evaporation no longer occurs, and that $P_r > 1$. However, the concept of a superheat no longer has meaning. Therefore, the expander inlet temperature can be specified directly as a model input. Furthermore, the pinch point can no longer be easily defined as the state-point '2' no longer exits. One way around this is to instead define the heat-source temperature drop (i.e. $T_{hi} - T_{ho}$) and vary this parameter as part of the system design. In this case, it is then necessary to construct the full isobar between state-points '2' and '3' and ensure the pinch point is not violated during the heat-addition process.

5.2   Working-Fluid Selection

The selection of a working fluid for a Rankine cycle system is dependent upon many factors, including thermodynamic performance, component performance, material compatibility, safety, environmental properties, availability and cost. A potential working fluid could be selected from a number of different groups, for example hydrocarbons (HC), hydrofluorocarbons (HFC), hydrofluoroolefins (HFO), siloxanes (SI) or could be a fluid such as water, ammonia or carbon dioxide. Chlorofluorocarbons (CFC) and hydrochlorofluorocarbons (HCFC) may also be suitable from a thermodynamic point of view, but their use has been phased out due their detrimental effect on the environment. Table 5.1 summarises working fluids commonly found within commercial ORC plants to give the reader a feel for typical working fluids.

Badr, Probert and O'Callaghan (1985) presented an early study working-fluid selection, and discussed the properties that a working fluid should ideally exhibit. It is quickly established that there is no unique working fluid which will satisfy all the desired criteria, and it is often up to the designer to select a fluid based on their own selection criteria. The desired properties discussed have been summarised in the following list. These same selection criteria have been reiterated more recently by a number of authors (Badr, Probert & O'Callaghan 1985; Maizza & Maizza 1996; Tchanche et al. 2011; Vélez et al. 2012; Rahbar et al. 2017).

- First and foremost, the working fluid must result in an optimal thermal cycle efficiency resulting in an optimal conversion of the input heat into power.
- The evaporation pressure should not be excessive to avoid high mechanical stress and expensive component design.
- The condensation pressure should be above atmospheric pressure to avoid the requirement of operating the condenser under a vacuum.
- The minimum ambient temperature should be above the fluids, triple point.

Table 5.1 Common ORC fluids as suggested by Colonna et al. (2015)

Fluid	Chemical Formula	Molecular Weight g/mol	Critical Temperature [°C]	Critical Pressure [bar]
Toluene	C_7H_8	92.1	318.6	41.26
Cyclopentane	C_5H_{10}	70.1	238.5	45.15
Isopentane	C_5H_{12}	72.1	187.2	33.78
Isobutane	C_4H_{10}	58.1	134.7	36.29
MDM[1]	$C_8H_{24}Si_3O_2$	236.5	290.9	14.15
MM[2]	$C_6H_{18}OSi_2$	162.4	245.5	19.39
PP1[3]	C_6F_{14}	338.0	182.2	19.23
R245fa[4]	$C_3H_3F_5$	134.0	154.0	36.51
R134a[5]	$C_2H_2F_4$	102.0	101.1	40.59

[1] Octamethyltrisiloxane   [2] Hexamethyldisiloxane
[3] Perfluoro-2-methylpentane   [4] 1,1,1,3,3-pentafluoropropane
[5] 1,1,1,2-tetrafluoroethate

- High latent heat of vaporisation and high thermal conductivity are advantageous to achieve high heat transfer rates within the heat exchangers.
- Low viscosity can help to reduce pressure drops within the system.
- High fluid density results in low volume flow rates, permitting small cycle components to be designed, whilst minimising pressure losses downstream of the expander.
- The slope of the fluids saturated vapour line should be close to vertical (see Figure 5.3). Fluids with a negative gradient require superheating to ensure the expansion finishes within the superheated region, whilst fluids with a positive gradient exit the expander with a large superheat requiring a large amount of precooling prior to condensation.
- The fluid should be non-corrosive and compatible with the materials used for the construction of the system.
- The fluid should be chemically stable within the operating range being considered.
- The fluid should be non-toxic and non-flammable.
- The fluid should have good lubrication properties.
- The fluid should be low cost.

Alongside the properties already listed, it is also important to consider the environmental properties of the working fluid. Due to the Montreal and Kyoto protocols a number of potential working fluids have already been banned, with others set to be phased out. A possible working fluid should therefore have a low environmental impact with a low global warming potential, a low atmospheric lifetime and a low ozone-depletion potential.

From a thermodynamic point of view, two thermodynamic properties that can be used to assess the feasibility of a particular fluid are the normal-boiling temperature

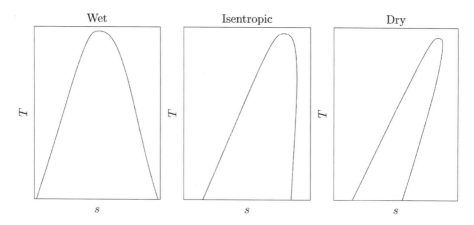

Figure 5.3 Classification of fluids according to the slope of their saturated-vapour dome. Fluids with negative, infinite and positive gradients are classified as 'wet', 'isentropic' and 'dry' fluids, respectively

T_b (determined at 1 bar) and the critical temperature T_{cr}, and these have been plotted in Figure 5.4. In general, there is a link between heat-source temperature and the optimal critical temperature of the working fluid (Chen, Goswami & Stefanakos 2010; White & Sayma 2018), and thus fluids with lower critical temperatures can will be suitable low-temperature heat sources, whilst fluids with high critical temperatures will be more suitable for higher-temperature heat sources. However, from Figure 5.4 it is observed that as the critical temperature increases, so does the normal-boiling temperature. As a consequence, working fluids with high critical temperatures will typically result in sub-atmospheric condensation pressures, which will increase system complexity. Alternatively, if sub-atmospheric conditions are to be avoided, it becomes necessary to increase the condensation temperature, which could thus have a negative effect on the cycle's thermodynamic performance.

In general, working-fluid selection criteria are introduced during a fluid-selection study, in which a group of known fluids, taken from a database such as NIST REF-PROP (Lemmon, Huber & McLinden 2013), is screened based on predefined criteria. Then, for the identified working fluids a parametric optimisation study is completed in which the ORC is optimised for each working fluid in turn and the results are compared.

5.3 Performance Evaluation of a Rankine Cycle

Having now introduced the reader to the key aspects of modelling of a Rankine cycle and having discussed working-fluid selection criteria, it is now appropriate to use the models developed to investigate the performance of Rankine cycles under different operating conditions.

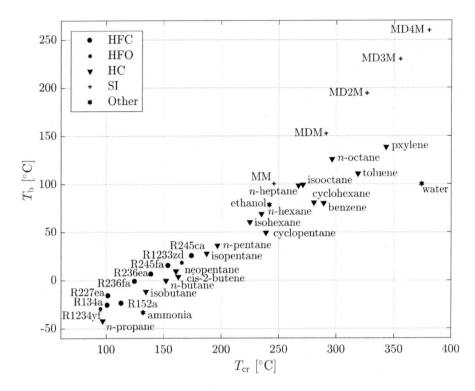

Figure 5.4 Potential ORC working fluids plotted in terms of their critical temperature T_{cr} and normal-boiling temperature T_b. The main groups considered at hydrofluorocarbons (HFC), hydrofluoroolefins (HFO), hydrocarbons (HC), siloxanes (SI) and others

5.3.1   Evaluation of a Simple Rankine Cycle

Referring back to the simple, subcritical, non-recuperated Rankine cycle model described in Section 5.1.2, it is apparent that the thermodynamic performance of the cycle, measured either in terms of the power output or the thermal efficiency, is a function of three thermodynamic variables: the condensation temperature T_1, the reduced pressure P_r and the amount of superheating ΔT_{sh}. Therefore, it is useful to conduct a number of simple parametric investigations to understand how these three parameters effect the performance of the cycle. For these investigations we define a heat source and consider an ORC operating with R245fa, which is a common working fluid used within ORC applications. The remainder of the assumptions are listed in Table 5.2.

The aim of the first investigation is to establish the effect of T_1 and P_r on cycle performance. For this investigation these two parameters were varied parametrically over a range of values and their impact on the thermal efficiency η_{th}, the net specific work w_n, the working-fluid mass-flow rate $\dot{m}_{wf}$ and the net power output $\dot{W}_n$ of a simple, subcritical, non-recuperated ORC system was determined. This analysis was completed assuming a fixed superheat of $\Delta T_{sh} = 10\ °C$. The results are shown in Figure 5.5.

Table 5.2 Assumptions made for the parametric investigation

ORC working fluid	-	R245fa	-
Heat-source temperature	T_{hi}	180	°C
Heat-source mass-flow rate	$\dot{m}_h$	1.0	kg/s
Heat-source specific-heat capacity	$c_{p,h}$	4200	J/(kg K)
Pump isentropic efficiency	η_p	0.7	-
Expander isentropic efficiency	η_e	0.8	-
Evaporator pinch point	PP_h	10	K

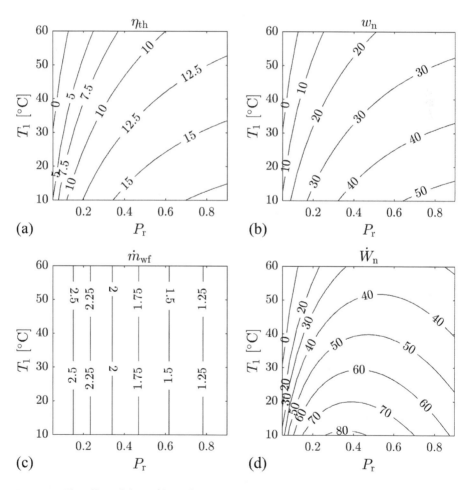

Figure 5.5 The effect of the condensation temperature T_1 and reduced evaporation pressure P_r on the thermal efficiency η_{th}, net specific work w_n, working-fluid mass-flow rate $\dot{m}_{wf}$ and the net power output $\dot{W}_n$ of an organic Rankine cycle operating under the conditions defined in Table 5.2. Results based on an assumed superheat of $\Delta T_{sh}=10$ K

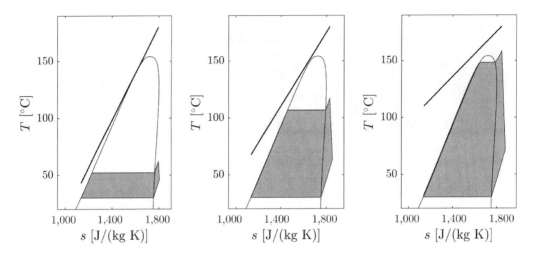

Figure 5.6 T–s plots and heat-source temperature profiles for three different cycles. From left to right: $P_r = 0.1, 0.4$ and 0.9

From Figure 5.5 it is apparent that η_{th} and w_n are directly proportional to each other and are maximised when T_1 is minimised and P_r is maximised. Since minimising T_1 also corresponds to minimising the condensation pressure, and maximising P_r corresponds to maximising the evaporation temperature, it follows that η_{th} and w_n are maximised when the ORC pressure ratio and the temperature difference between the condensation and evaporation processes are maximised.

In terms of the working-fluid mass-flow rate it is observed that, for a fixed superheat and evaporation pressure, $\dot{m}_{wf}$ is independent of T_1. The reason for this is because $\dot{m}_{wf}$ is determined from an energy balance applied to the evaporator that considers only the two-phase evaporation and superheating processes. These processes are dependent only on the evaporation pressure, and therefore T_1 has no effect on $\dot{m}_{wf}$. On the other hand, it is observed that $\dot{m}_{wf}$ reduces as P_r increases, and this is due to the interaction between the heat source and the working fluid as the reduced evaporation pressure increases. To investigate this further, the temperature–entropy (T–s) diagrams for three cycles at different reduced pressures are plotted in Figure 5.6, alongside the heat-source temperature profiles. As P_r increases, the evaporation temperature increases, which results in, alongside the imposed evaporator pinch point, a smaller heat-source temperature drop ΔT_h. Since $Q_h = \dot{m}_h c_{p,h} \Delta T_h$, a smaller ΔT_h corresponds to the ORC absorbing less of the available heat from the heat source, which in turn reduces the working-fluid mass-flow rate.

Ultimately, the coupled effect of w_n increasing with P_r and $\dot{m}_{wf}$ reducing with P_r is that at a defined condensation temperature there exists an optimal P_r at which the net power output, $\dot{W}_n = \dot{m}_{wf} w_n$, is maximised. This is observed in Figure 5.5(d). Comparing the results for $\dot{W}_n$ with the results for η_{th} in Figure 5.5, an important characteristic of a Rankine cycle is identified: the optimal cycle that maximises the power output from the system is the not the same system that optimises the thermal efficiency. The result of this is that the design of the system is strongly linked to the application.

For example, an ORC system designed to convert solar heat into electricity should be designed to maximise efficiency since the heat source leaving the ORC evaporator is returned to the solar collector to be reheated. On the other hand, for a waste-heat recovery application, in which the waste-heat is free and is typically rejected to the ambient surroundings, the system should be designed to maximise the power output.

In Figure 5.7 the results from another parametric investigation are shown. In this investigation, P_r and ΔT_{sh} are varied parametrically, whilst T_1 is fixed at 20 °C. Firstly, it is observed that for a fixed P_r, ΔT_{sh} has a minimal effect on η_{th}. However, the results also show that at high reduced pressures, a larger superheating can result in a higher w_n. In terms of $\dot{m}_{wf}$, it is again observed that increasing P_r corresponds to a reduction in the working-fluid mass-flow rate. However, an increasing superheat is also detrimental to $\dot{m}_{wf}$. This is because an increasing superheat means that the working fluid is heated to a higher temperature, which in turn means it must absorb more heat from the heat source. For a finite heat supply this must result in a reduced mass-flow rate. Finally, $\dot{W}_n$ is plotted in Figure 5.7(d). Ultimately, the results in this plot show that the power output is maximised when the superheating is reduced, and therefore there is no thermodynamic benefit in superheating the working fluid before expansion.

Alongside T_1, P_r and ΔT_{sh}, another important parameter that affects the performance of a Rankine cycle is the evaporator pinch point PP_h. In Figure 5.8, the effect of PP_h on the system has been investigated. For a range of pinch points T_1, P_r and ΔT_{sh} were optimised to generate the maximum power from the cycle, based on the same assumptions listed in Table 5.2. In addition to evaluating $\dot{W}_n$, the effect of the pinch point on the evaporator was also evaluated by considering the product of the overall heat-transfer coefficient U and heat-transfer area A for the preheating, evaporation and superheating processes, and adding these together. Therefore, the parameter plotted in Figure 5.8 is defined as:

$$UA = \frac{\dot{Q}_{ph}}{\Delta T_{log,ph}} + \frac{\dot{Q}_{ev}}{\Delta T_{log,ev}} + \frac{\dot{Q}_{sh}}{\Delta T_{log,sh}}, \qquad (5.43)$$

where $\dot{Q}_{ph}$, $\dot{Q}_{ev}$ and $\dot{Q}_{sh}$ are the heat-transfer rates in the preheater, evaporator and superheater, respectively and $\Delta T_{log,ph}$, $\Delta T_{log,ev}$ and $\Delta T_{log,sh}$ are the log-mean temperature differences in the preheater, evaporator and superheater, respectively. Defined in this way UA can be considered a size factor, with larger values of UA corresponding to larger heat exchangers.

Ultimately, from the results in Figure 5.8, it is observed that reducing the evaporator pinch point leads to an increase in the power output. However, reducing the pinch point also leads to an increase in UA, corresponding to larger heat-transfer areas. Furthermore, as the pinch point reduces below 10 °C, UA starts to increase more rapidly, and eventually, as $PP_h \rightarrow 0$, $UA \rightarrow \infty$. Therefore, the evaporator pinch point represent a trade-off between thermodynamic performance, and the size and cost of the heat exchangers.

The trade-off between thermodynamic performance and heat-exchanger size is one that has been discussed in detail within the literature. In particular, multiple authors have coupled thermodynamic cycle analysis to heat-exchanger sizing models in order to conduct multi-objective optimisation. These aspects are discussed in more detail in Sections 5.4.4 and 5.5.3.

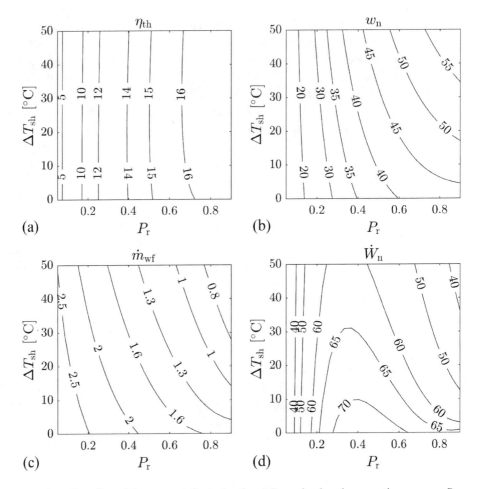

Figure 5.7 The effect of the amount of superheating ΔT_{sh} and reduced evaporation pressure P_r on the thermal efficiency η_{th}, net specific work w_n, working-fluid mass-flow rate $\dot{m}_{wf}$ and the net power output $\dot{W}_n$ of an organic Rankine cycle operating under the conditions defined in Table 5.2. Results based on an assumed condensation temperature of $T_1 = 20\,°C$

5.3.2 A Comparison of Water, R245fa and Cyclopentane

In addition to evaluating how the performance of a Rankine cycle is affected by the main operating parameters, it is also interesting to investigate how the performance of a Rankine cycle depends on the working fluid selected. This is useful to demonstrate how different working fluids are optimal for different heat-source temperatures, and to also show why it is not always suitable to operate a steam Rankine cycle.

For this study we consider three different working fluids, namely: water, R245fa and cyclopentane. R245fa and cyclopentane are both commonly used working fluids in ORC systems (Table 5.1) and have critical temperatures of 134.0 °C and 238.5 °C, respectively. Therefore, they are suitable working fluids for relatively low-temperature applications and slightly higher-temperature applications, respectively.

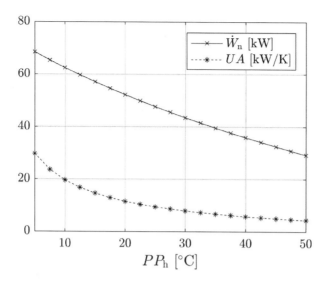

Figure 5.8 The effect of the evaporator pinch point PP_h on the net power output $\dot{W}_n$ and the heat exchanger size parameter UA defined by Equation 5.43

For each working fluid the optimal simple, subcritical, non-recuperated cycle was obtained for a range of heat-source temperatures. Each optimal cycle was obtained by optimising the reduced evaporation pressure P_r and the amount of superheating ΔT_{sh}, assuming a fixed heat-source heat capacity rate ($\dot{m}_h c_{p,h} = 4{,}200$ W/K), fixed component efficiencies ($\eta_p = 0.7$, $\eta_e = 0.8$) and a fixed evaporator pinch point ($PP_h = 10\ °C$). These are the same assumptions listed in Table 5.2. For the R245fa case, the condensation temperature was set to $T_1 = 30\ °C$, which corresponds to a condensation pressure of $P_1 = 1.77$ bar absolute. The condensation pressures for both cyclopentane and water were set to 1 bar absolute to avoid sub-atmospheric condensation, which corresponds to $T_1 = 49.3\ °C$ and $T_1 = 100\ °C$ for cyclopentane and water, respectively. Finally, a constraint was applied to the expander outlet conditions to ensure that expansion remained within the superheated region. The results from the optimisations are shown in Figure 5.9.

First and foremost, it is clear that for all the heat-source temperatures considered, the net power output from the ORC systems is significantly higher than that of the steam Rankine cycle, affirming that ORC systems are the optimal choice within this range of heat-source temperature. In terms of working-fluid selection, it appears that, in order to maximise $\dot{W}_n$, R245fa should be selected when $T_{hi} < 250\ °C$, and cyclopentane should be selected for $T_{hi} > 250\ °C$. In terms of the cycle thermal efficiency η_{th}, the optimal cycles range between $\eta_{th} = 6\%$ for R245fa at $T_{hi} = 100\ °C$ and $\eta_{th} = 20\%$ for cyclopentane at $T_{hi} = 300\ °C$. The Carnot efficiencies for these heat-source temperatures, based on an assumed heat sink at $30\ °C$, are 18.8% and 47.1%, respectively, and therefore the two optimal cycles at these two heat-source temperatures achieve approximately 33% and 42% of the Carnot efficiency, respectively.

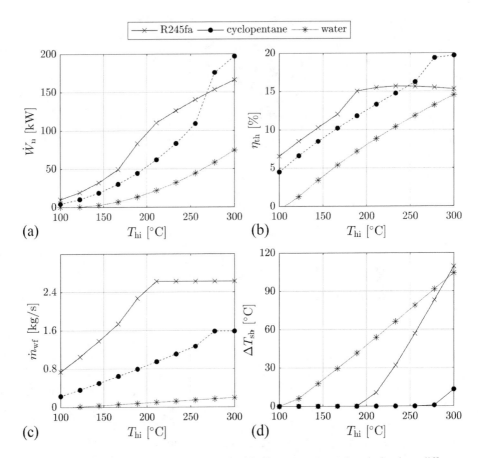

Figure 5.9 The performance of the optimal subcritical, non-recuperated cycle for three different working fluids at different heat-source temperatures T_{hi}. Results are plotted in terms of the net power output $\dot{W}_n$, the thermal efficiency η_n, the working-fluid mass-flow rate $\dot{m}_{wf}$ and the amount of superheating ΔT_{sh}

When evaluating how $\dot{W}_n$ and η_{th} vary as the heat-source temperature increases for R245fa and cyclopentane, it is observed that the results do not form continuous curves, but instead are discontinuous. This occurs because the evaporation pressure is constrained to ensure that the cycle operates in the subcritical region. At low heat-source temperatures the optimal evaporation pressure is below the maximum allowed pressure and therefore the optimal amount of superheating tends to zero. However, as the heat-source temperature increases, so does the evaporation temperature, which also correlates to a higher evaporation pressure. Eventually, as the heat-source temperature is increased further, the evaporation pressure tends towards the maximum, and then the only way for the working fluid to absorb more heat from the waste-heat stream is via superheating. It is this transition from a cycle with an evaporation pressure that is lower than the maximum allowed pressure and a superheat of zero to a cycle operating at the maximum pressure with superheating that causes the discontinuous behaviour.

This behaviour is also observed in Figure 5.9(d) for the R245fa and cyclopentane cases. Furthermore, it should be noted that when the evaporation pressure is on the upper bound, and higher heat-source temperatures can only be utilised by increasing the superheating, the ORC pressure ratio does not change. As a result of this the specific work, which is proportional to the pressure ratio, no longer increases as rapidly, resulting in the thermal efficiency flattening off, as observed for R245fa when $T_{hi} > 200\ °C$. It is also observed that the working-fluid mass-flow rate flattens off at high heat-source temperatures. This is because operating a cycle at the maximum evaporation pressure with a high amount of superheating results in most of the heat transfer from the heat source to the working fluid occurring within the preheating and superheating regions. This allows a close thermal match between the heat source and working fluid to be obtained, and results in two evaporator pinch points: one at the inlet to the evaporator (i.e. $T_{ho} - T_2 = PP_h$) and the other at the start of evaporation (i.e. $T_{hp} - T_{2'} = PP_h$). This, in turn, fixes $\dot{m}_{wf}$.

Finally, it is worth commenting on the results for the optimal Rankine cycles operating with water, even though it has already been shown that the net power output and thermal efficiency from these cycles is lower than the corresponding cycles operating with R245fa and cyclopentane. Firstly, it is observed that the steam Rankine cycles correspond to much lower working-fluid mass-flow rates than the ORC systems. Furthermore, since the relative reduction in $\dot{m}_{wf}$ when comparing R245fa and water is much larger than the relative reduction $\dot{W}_n$, it follows that the expander specific work, $w_e = h_3 - h_2$, for the steam Rankine cycle is larger. A large specific work can be problematic to expander design, which is further compounded by the low flow rates, particularly for small-scale systems. Referring to Figure 5.9(d) it is also observed that steam Rankine cycles all correspond to a large amount of superheating. This is because water is a wet fluid, which means the slope of the saturated-vapour line is negative. Therefore, to ensure the expansion process occurs within the superheated region, it is necessary to apply a large superheat to the working fluid. This requirement lowers the cycle pressure ratio, and results in a lower thermal efficiency and lower power output.

5.3.3   The Performance of Alternative Cycles

Following on the previous analysis in Section 5.3.1, it is also interesting to investigate how alternative cycle architectures could be used to enhance the thermodynamic performance of the system. Therefore, in the following subsections, the performance of the simple, subcritical, non-recuperated cycles identified previously will be compared to recuperated cycles, working-fluid mixtures, partially evaporated cycles and supercritical cycles.

Recuperated Cycles

To investigate the potential benefits of installing a recuperator within a Rankine cycle another parametric investigation can be completed comparing the performance of the cycle with and without recuperation. For this analysis $T_1 = 30\ °C$, and it is assumed that expansion occurs from a saturated-vapour state (i.e. $\Delta T_{sh} = 0\ °C$), and that the

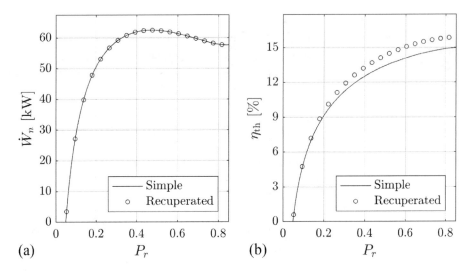

Figure 5.10 Comparison between a simple and recuperated Rankine cycle in terms of the net power output $\dot{W}_n$ and thermal efficiency η_{th}. Results obtained for the assumptions given in Table 5.2 assuming $T_1 = 30\,^\circ\text{C}$, $\Delta T_{sh} = 0\,^\circ\text{C}$ and $\epsilon_r = 0.7$

recuperator effectiveness is $\epsilon_r = 0.7$. Then, for the assumptions listed in Table 5.2 the reduced evaporation pressure can be varied parametrically. The results in Figure 5.10 compare the net power output and thermal efficiency of a Rankine cycle operating with and without recuperator.

From Figure 5.10(a) it is clear that the recuperated cycle generates exactly the same amount of power as the simple cycle. The reason for this is because the net specific work is the same for both cycles since the pump and expander inlet and outlet conditions are the same. Furthermore, the working-fluid mass-flow rate is dependent only on the energy balance applied to the evaporation and superheating processes, and since this energy balance is also the same in both cycles, the net power output does not change. However, when comparing the thermal efficiency it is observed that installing a recuperator leads to a higher efficiency. This is because the net specific work does not change, but the specific thermal input required to preheat the working fluid to a saturated liquid (i.e. $h_{2'} - h_2$) is less since the temperature of the working fluid at the evaporator inlet is higher. Consequently, this means that the heat-source outlet temperature T_{ho} is also increased, which means less heat is absorbed from the heat source.

Ultimately, for renewable applications such as a solar Rankine cycle where the heat-source outlet is returned to the solar collector, a recuperator can generally be used to beneficial effect to improve the thermal efficiency of the system. On the other hand, for a waste-heat recovery application, when one wants to generate the maximum power from a particular waste-heat stream, there is generally no benefit in installing a recuperator. This conclusion has been supported by the conclusions from other researchers working within the field (Dai, Wang & Gao 2009; Quoilin et al. 2011). The one caveat to this is in waste-heat recovery applications where there is a limitation on the temperature to which the waste-heat stream can be cooled (e.g. to avoid condensation), or where there are constraints on the cycle operating parameters. For example, the findings from Oyewunmi et al. (2017) suggest that recuperators are not necessary for an unconstrained application, but when constraints on the minimum or maximum operating pressures are introduced, recuperators become beneficial.

Working-Fluid Mixtures

The aim of the next parametric investigation is to demonstrate the potential thermodynamic benefit in operating a Rankine cycle with a mixture. The advantage of using a working-fluid mixture results from non-isothermal evaporation and condensation processes, which in turn reduces irreversibility in the heat addition and heat rejection processes in the evaporator and condenser, respectively. Since the effect of using a working-fluid mixture impacts both the evaporation and condensation processes, to conduct a fair investigation into using working-fluid mixtures it is necessary to also define the heat sink, which has been defined with the following parameters: $T_{ci} = 15\,°C$, $\dot{m}_c = 5$ kg/s and $c_{p,c} = 4{,}200$ J/(kg K). Furthermore, the minimum allowable condenser pinch point has been defined as $PP_c = 5\,°C$. For this heat sink, alongside the parameters defined in Table 5.2, a parametric investigation considering a working-fluid composed of a mixture of n-pentane and n-hexane has been conducted. In this study, the mass fraction of n-pentane was parametrically varied between 0 and 1, and for each mass fraction an optimisation was completed in which T_1, P_r and ΔT_{sh} were optimised in order to generate the maximum net power output. The results from this analysis are shown in Figure 5.11.

Ultimately, what is apparent from Figure 5.11 is that a Rankine cycle operating with a mixture of n-pentane and n-hexane is capable of generating more power than a Rankine cycle operating with only n-pentane or n-hexane. Therefore, it can be concluded that the use of a mixture can allow a Rankine cycle system to extract more power from a given waste-heat source. However, given the large number of working fluids available, and the even larger number of potential working-fluid pairs that could be selected with varying mass fractions, the challenge then becomes the identification of an optimal working-fluid mixture for a particular application.

With reference to some published works, Angelino and Colonna (1998) present an early work exploring the potential of mixtures for ORC systems. The authors note that mixtures are particularly advantageous in applications where the heat source or heat sink are expected to exhibit significant temperature differences. The authors of that study also note that some effort is required to identify suitable mixtures that exhibit a suitable temperature glide. In terms of some latter studies, Lecompte et al. (2014) investigated mixtures for heat-source temperatures ranging between 120 and 160 °C and found that second law efficiencies were improved in the range of 7% and 14% when using mixtures, whilst Andreasen et al. (2016) found that for a 90 °C heat source, a mixture could result in 3.2% more power than a pure fluid for the same initial investment cost. On the other hand, the study by Oyewunmi et al. (2016) suggests that although mixtures result in higher power outputs than pure fluids, pure fluids could result in around 14% lower costs per unit power, owing to mixtures generally

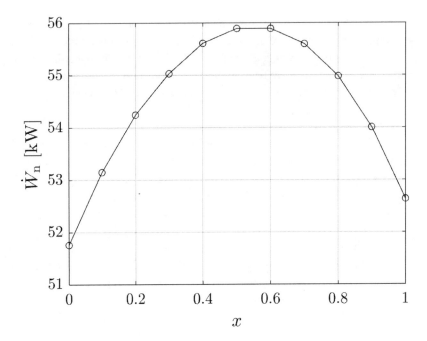

Figure 5.11 The effect of the mass fraction of n-pentane x in a mixture of n-pentane and n-heptane on the maximum net power output from a Rankine cycle

requiring larger heat exchangers. Ultimately, these studies demonstrate the potential of mixtures to enhance thermodynamic performance, but suggest that care must be given to consider the effect of the mixture on component design.

Trilateral or Partially Evaporated

The penultimate investigation completed as part of this section considers the use of partially evaporated cycles, in which the working fluid is expanded from a two-phase state. To investigate this type of system, and to compare this cycle to a simple Rankine cycle in which the working fluid is expander from a saturated-vapour or a superheated state, it is useful to introduce a single parameter z which is capable of describing both types of cycle, and can range between 0 and 2. When $0 \leq z \leq 1$, the working fluid is assumed to expand from a two-phase state and z is equal to the expander-inlet vapour quality. However, when $1 < z \leq 2$, the working fluid expands from a superheated state and the amount of superheating ΔT_{sh} is given by:

$$\Delta T_{sh} = (z - 1)(T_{hi} - T_{3'}), \tag{5.44}$$

where T_{hi} is the heat-source inlet temperature and $T_{3'}$ is the saturated-vapour temperature. It should be noted that when $z = 2$, $T_{hi} = T_{3'}$, which is a violation of the imposed evaporator pinch point. Therefore, within the pinch, constraint $(T_{hi} - T_{3'} > PP_{h,min})$ will always result in cycles where $z < 2$.

For the same assumptions defined previously (Table 5.2), and a fixed condensation temperature of $T_1 = 30\ °C$, a parametric investigation of P_r and z can be completed,

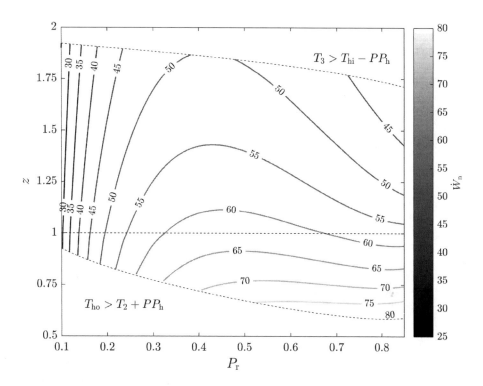

Figure 5.12 Comparison between partially evaporated cycles ($z \leq 1$) and simple cycles ($z > 1$) at different reduced evaporation pressures P_r. Results obtained for the assumptions given in Table 5.2 assuming $T_1 = 30\,^\circ\text{C}$

and the effect on the net power output from the system can be investigated. The results from such an analysis are given in Figure 5.12. For a given P_r, the limits for z are determined to ensure that the imposed pinch-point constraint is not violated within the evaporator.

From Figure 5.12 it is observed that at each reduced evaporation pressure the maximum net power output is obtained at the minimum value of z that does not violate the pinch-point heat-source outlet pinch-point (i.e. $T_{ho} > T_2 + PP_h$). Therefore, it is quite apparent that a partially evaporated cycle can have a significant advantage over the simple superheated cycle by allowing the heat source to be cooled down to a much lower temperature, therefore resulting in the working fluid absorbing more heat from the heat source, which in turn leads to a higher power output. When comparing the optimal simple ($z = 1$) and partially evaporated cycles ($z \leq 1$) in Figure 5.12, it is observed that the optimal power outputs are around 60 and 80 kW, respectively, and therefore in this instance, operating a partially evaporated cycle could lead to an approximate 33% increase in the $\dot{W}_n$. It is however noted that the optimal partially evaporated cycle has a higher reduced evaporation pressure ($P_r = 0.85$) compared to the simple cycle ($P_r = 0.5$).

A few notable studies from the literature include the work of Smith (Smith 1993; Smith & da Silva 1994; Smith, Stosic & Aldis 1996), who has studied the trilateral

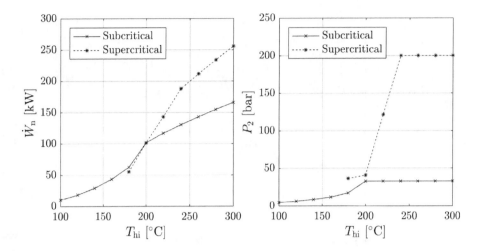

Figure 5.13 Comparison between the performance of an optimised subcritical and supercritical ORC operating with R245fa at different heat-source temperature T_{hi}. Results are shown in terms of the net power output $\dot{W}_n$ and the maximum operating pressure P_2

flash cycle in detail. One study suggested that for recovering heat below 200 °C the net power output from these cycles could be between 10% and 80% greater than a simple Rankine cycle (Smith 1993). More recently, Fischer (2011) compared trilateral and ORC systems under five different scenarios for the heat-source and heat-sink conditions. The results suggest that for heat-source temperatures between 350 and 220 °C , the trilateral cycle could produce between 14% and 20% more power, but this increases to 30% for a heat-source temperature of 150 °C.

Supercritical

The final investigation considers the potential performance benefit in operating supercritical Rankine cycle, in which the working fluid is pressurised to a pressure greater than the critical pressure and the two-phase evaporation process is removed. For this study we reconsider the optimisation completed in Section 5.3.2, in which an ORC operating with R245fa was optimised for a range of heat-source temperatures. Using the same values for $\dot{m}_h$, $c_{p,h}$, η_p, η_e, PP_h and T_1, this same optimisation has been completed again, but considering a supercritical ORC operating with the same working fluid. In this case, the reduced evaporation pressure P_r, expander-inlet temperature T_3 and heat-source outlet temperature T_{ho} are optimised to generate the maximum power from the heat source. Constraints are applied to ensure that T_3 is sufficiently large such that expansion occurs within the superheated region, whilst the maximum operating pressure is set to 200 bar absolute. A comparison between the results for the subcritical and supercritical cycles is shown in Figure 5.13.

Figure 5.13 shows that there could be a significant benefit in operating a supercritical cycle with R245fa as the heat-source temperature exceeds 200 °C, compared to a subcritical ORC operating with the sameworking fluid. Doing so results in a much higher pressure ratio, which coupled to a better thermal match between the heat source and working fluid results in a higher net power output from the system. However, the major drawback is the excessively high operating pressure which introduces more design complexity to ensure safety regulations are met. This would inevitably have cost implications.

It is worth noting that although it appears that a supercritical ORC operating with R245fa could be an effective solution for heat-source temperatures exceeding 200 °C, there could exist an alternative working fluid which can achieve similar thermodynamic performance whilst operating a subcritical cycle. Therefore, the optimal system can only really be identified following a thorough comparison of different working-fluids and cycle architectures. Along this vein, Chen, Goswami and Stefanakos (2010) attempted to assess a range of different working fluids and categorise them according to whether they are best suitable for subcritical or supercritical cycles. Some further examples of studies addressing the thermodynamic optimisation of supercritical ORC systems can be found in the works of Schuster, Karellas and Aumann (2010) and Le et al. (2014).

5.3.4   Summary

Ultimately, from the analysis presented in this section, it is possible to summarise a number of key conclusions which can help in identifying the optimal cycle architecture for a particular application. These are as follows, and are presented from the point of view of a thermodynamic optimum:

- In general, it can be observed that for a subcritical, non-recuperated cycle:
  - The optimal condensation should be the minimum condensation temperature that can be maintained by the heat sink, without violating the condenser pinch constraint.
  - Maximising the evaporation pressure will result in the highest thermal efficiency. However, when considering the net power output there is often an optimal evaporation pressure at which maximum power will be produced.
  - As long as the optimal evaporation pressure is less than the maximum allowable pressure the optimal amount of superheating will tend to zero, indicating there is no thermodynamic benefit in superheating the working fluid.
  - As the heat-source temperature increases and the evaporation pressure tends towards the maximum, the working fluid should be superheated. This is because the maximum evaporation temperature also becomes fixed, meaning that the only way to absorb more heat into the working fluid is via superheating.
  - Minimising the evaporator pinch point will always facilitate the working-fluid absorbing the maximum amount of heat, corresponding to the best thermodynamic performance. However, this comes at the cost of larger heat exchangers.

- It has been shown that for heat-source temperatures between 100 and 300 °C a steam Rankine cycle cannot outperform and ORC.

- Unless there is a use for the waste heat leaving the ORC evaporator, or the heat-source outlet temperature is constrained by a downstream process, there is no benefit in installing a recuperator within an ORC intended for waste-heat recovery applications.
- Alternative cycle architectures such as working-fluid mixtures, partially evaporated and supercritical cycles, can improve on the net power output from a subcritical, non-recuperated cycle.

Despite being able to pull this list of guidelines together, it is worthwhile to point out that given the large number of working fluids available, the range of cycle architectures that are possible, and the wide range of potential waste-heat recovery applications, there is no substitute for conducting a comprehensive study for a particular waste-heat stream. Such a study should comprehensively evaluate and compare different working fluids and cycle architectures to identify the correct choice for a given application.

Furthermore, whilst the thermodynamic benefit of novel architectures has been discussed as part of this section, it is also worth noting that these cycles do introduce technical challenges that need to be overcome before the potential performance benefits can be realised.

- All of the alternative cycles aim at improving the thermal match between the heat source and working fluid, which in turn results in the cycle absorbing more heat from the heat source. This in turn allows the system to generate more power. However, this improved thermal match will come at the cost of larger heat exchangers.
- For working-fluid mixtures the problem mentioned in the previous point can be further compounded by lower heat-transfer coefficients than for pure fluids.
- The success of partially evaporated cycles requires the development of suitable two-phase expanders which are currently not commercially available.
- The high operating pressures within supercritical cycles place more restrictive constraints on system design, and also require expanders design for high-expansion ratios.

5.4   System Components

Following from working-fluid selection and the identification of the optimal cycle architecture and operating conditions, it is necessary to consider component selection. Predominantly, a Rankine cycle system is composed of an expander, heat exchangers and a pump; each of these technologies will now be introduced.

5.4.1   Expander

The expander is arguably the most critical component with a Rankine cycle, responsible for generating the mechanical shaft power that can be converted into electricity

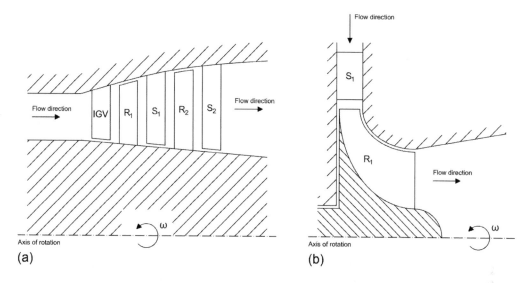

Figure 5.14 Schematic of a multi-stage axial turbine (a) and a radial turbine (b). The letters R_n and S_n correspond to the rotor and stator of the nth stage, whilst IGV stands for inlet-guide vanes

by coupling it to a generator. Primarily, an expander will fall into one of two categories, namely: turboexpanders or volumetric expanders. It is not the purpose of this book to delve into the topic of designing or selecting an expander for an ORC system in depth, but rather to provide a brief overview of the operating principles and some general sizing guidelines. For more details, Alshammari, Usman and Pesyridis (2018) provide a recent review on the status of ORC expander technology, while Persico and Pini (2017) provide more details relating to turboexpanders, and the works of Smith, Stosic and Kovacevic (2014) and Lemort and Legros (2017) address topics related to volumetric expanders.

Turboexpanders

A turboexpander is a machine where the dynamic action of a rotating rotor accelerates the fluid passing through it, changing its direction and thus causing a reduction in the stagnation enthalpy. This change in energy produces torque on the shaft and generates useful mechanical power (Dixon 2005). Typically, a turboexpander is formed from a stationary stator and the rotating rotor. The purpose of the stator is to accelerate the flow and to deliver the flow to the rotor with a large absolute tangential velocity. Then, as the flow passes through the rotor, the absolute tangential velocity is reduced, and this reduction in angular momentum corresponds to a torque on the rotor shaft. The two most common types of turboexpander are the axial turbine and the radial turbine, and a schematic of these expander types is given in Figure 5.14.

In the axial design, the flow enters from the left and moves from left to right, parallel to the axis of rotation, albeit with a varying tangential velocity. Throughout the expansion process the flow maintains an axial velocity component, but does not have a radial velocity component. One advantage of this geometry is that it allows multiple expansion stages to be easily mounted in series on the same shaft. Each stage consists of one rotor stage and one state stage, and this is indicated in Figure 5.14, where R_n and S_n correspond to rotor and stator of stage n, respectively. Referring to the axial turbine in Figure 5.14, the blades of the first stationary stage are typically referred to as the inlet-guide vanes.

In comparison, in the radial turbine the flow enters the rotor from above, and therefore is composed only of radial and tangential velocity components. Then, as the flow passes through the rotor it is turned $90°$, leaving the rotor with only axial and tangential velocity components. The geometry of this design means that it is much harder to achieve multi-stage expansion as multiple stages can no longer be mounted on the same shaft. However, because there is a reduction in radius between the inlet and outlet, and therefore a larger change in angular momentum, the radial turbine is capable of generating a higher work output over a single stage than the axial design.

In terms of performance, an axial turbine can generally achieve a high turbine efficiency over a wider range of rotational speed than a radial turbine. However, over the range of speeds where radial turbine achieves a high efficiency, it can be hard to find a decisive advantage in either design (Dixon 2005). However, for smaller-scale applications radial turbines are typically favoured over the axial design, owing to them being compact, having good manufacturability, lightweight construction, can achieve high efficiencies over a single-stage expansion and provide a robust design. The disadvantages associated with turboexpanders largely relate to small-scale turbines. As the power rating reduces, turbines can become expensive due to very high rotational speeds which may require high-ratio gearboxes or high-speed generators. Furthermore, as the turbine size reduces, the relative clearances increase, resulting in high tip clearance losses. This, alongside increased viscous losses, has a detrimental effect on turbine efficiency.

Volumetric Expanders

Volumetric expanders do not rely on the velocity of the fluid but instead expand the fluid through the cyclic change of volume of an expansion chamber. The most common types of volumetric expanders are screw, scroll and reciprocating-piston expanders.

A screw expander consists of two meshing helical rotors, as shown in Figure 5.15. As these rotors rotate, a number of working fluid chambers are created between the rotors and the expander casing. The high-pressure, high-temperature gas enters at one end of the expander, and as the rotor rotates the volume of the working fluid chamber continually increases, allowing the fluid to expand, producing mechanical power. The fluid is then discharged at the other side of the screw expander.

A scroll expander is formed from two identical scroll wraps which are constructed from an involute of a circle. After positioning the first scroll, the second is rotated by $180°$, and then offset such that the two scrolls come into contact at a number of locations, thus forming a number of working-fluid chambers. The second scroll orbitsaround a central point which causes these contact points to move, creating a

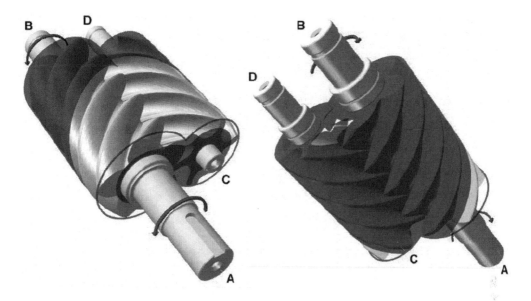

Figure 5.15 Three-dimensional model of the rotors from a twin-screw expander. Taken from Smith, Stosic and Kovacevic (2005)

Figure 5.16 Schematic of the operating principle of a scroll expander

variation in the chamber volume. The construction of a scroll expander is demonstrated in Figure 5.16, which is also used to demonstrate how the expansion process occurs. Starting on the left of this figure, the high-pressure, high-temperature working fluid enters at the centre of the scrolls. As the second scroll moves anticlockwise, more fluid is drawn into the scroll chamber until a full revolution is completed, at which point the chamber volume is shut-off from the inlet supply (this is demonstrated by the central plot in Figure 5.16). At this point the expansion of the fluid begins, and as the second scroll continues to rotate, the chamber volume increases, thus expanding the fluid to lower pressure and generating mechanical power. Then finally, as the rotation continues, the outlet port opens and the fluid is discharged from the scroll.

Finally, in a reciprocating-piston expander, expansion is achieved by the cyclic motion of a piston moving up and down in a cylinder. A schematic of a reciprocating-piston expander in given in Figure 5.17. At the top (top-dead centre) and the bottom of the piston strokes (bottom-dead centre), the volumes within the cylinder are at their minimum and maximum, respectively. Therefore, as the piston moves from top-dead

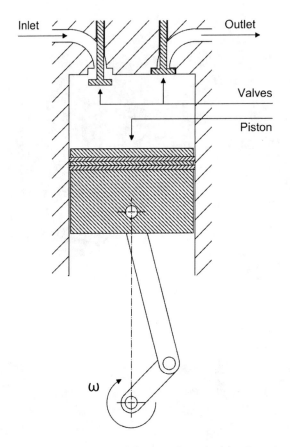

Figure 5.17 Schematic of the operating principle of a reciprocating-piston expander

centre to bottom-dead centre, a mass of fluid occupying the cylinder volume can expand, thereby producing mechanical power. Over one complete revolution of the shaft, during which the piston moves from top-dead centre to bottom-dead centre, and then back to top-dead centre, there are four different processes that occur. These are intake, expansion, discharge and re-compression. These four processes are controlled by inlet and outlet valves that control the intake and discharge processes.

Historically, screw and scroll expanders have been obtained by the reverse operation of existing screw and scroll compressors. Reciprocating-piston technology is already widely used, for example within the internal combustion engine and in steam engines; however, the application of reciprocating-piston expanders for ORC applications is a newer area of research. In general, volumetric expanders are generally preferred for smaller-scale applications, and offer a simple, low-cost alternative to turboexpanders, which generally have a low rotational speed, a low part count, are reliable, can tolerate two-phase conditions and can have improved off-design performance compared to turboexpanders. However, the general disadvantage associated with volumetric machines is that the expander efficiency can be lower than turboexpanders, which will impact the overall ORC performance.

5.4.2 Expanders in Commercial Systems

As discussed in Section 3.2.8, there are a number of commercial ORC systems available, and the expanders used in these technologies were given in Table 3.2. References for these companies are also given in Appendix A.

Overall it is observed that for large-scale applications turboexpanders are generally the preferred expander technology. Much of the commercialisation of ORC technology can be attributed to ORMAT and Turboden, and both of these companies utilise a multi-stage axial turbine in their ORC systems. There are also a number of companies utilising radial turbines, for example Atlas Copco, Calnetix, GE and Triogen. Exergy also utilise a turboexpander in their ORC systems, but their expander is a radial-outflow expander in which the working fluid enters at the centre of the expander and moves outwards in the radial direction, often across multiple stages. Maxxtec also utilise a turboexpander, but the architecture of this turbine is not known. As the system size reduces to a few hundred kW, there is interest in using volumetric expanders. For example, BEP and Electratherm have both commercialised ORC systems that utilise screw expanders. At the an even smaller scale of a kW up to a few tens of kW, there are systems available operating with scroll expanders, for example Eneftech. However, it is noted that small-scale systems are typically expensive bespoke systems, or are not installed in the large numbers required to confirm a widespread adoption of the technology. However, it is noted that the developed new expander technologies for small-scale ORC systems remains an active area of research, with screw, scroll and reciprocating-piston expanders all being investigated.

5.4.3 Expander Modelling and Selection Guidelines

When considering an expander for a particular waste-heat recovery applications there are a number of aspects that should be considered, and these affect the expander selected. The main two parameters are the pressure ratio (or volume ratio) across the expander and the volumetric flow rate.

The pressure and volume ratios (defined as the ratio of the expander inlet and outlet densities, i.e. ρ_3/ρ_4) are closely related, and are both output from a thermodynamic optimisation for a given waste-heat stream. In general, as the heat-source temperature increases, the optimal pressure ratio and volume ratio across the expander increases. High-volume ratios introduce more complexity into the expander design process, either through requiring large built-in volume ratios for volumetric expanders or through large changes in the required flow area within a turboexpander. In some cases, it may not be possible to achieve such high-volume ratios using a single volumetric expander, and therefore it could become necessary to consider multi-stage expansion. As a general observation, scroll and screw expanders are generally suitable for volume ratios in the order of 10 and below, whilst reciprocating-piston expanders are being investigated for much higher volume ratios. In terms of turboexpanders, large changes in the flow area can result in a requirement for very small blade heights at the rotor inlet, which increase losses within the turbine.

The volumetric-flow rate is very closely related to the size rating for the expander. For example, for a fixed heat-source temperature there will exist an optimal pressure ratio. However, as the heat-source mass-flow rate increases, the working-fluid mass-flow rate and therefore volumetric-flow rate will directly scale up. As discussed previously, turboexpanders are generally selected for power outputs exceeding a few hundred kilowatts, and volumetric expanders can be considered as the power output reduces below this. As another general trend regarding volumetric expanders, it is observed that scroll expanders tend to be most suitable within the power range of a kilowatt up to tens of kilowatts, whilst screw expanders are suitable within the range of a few tens of kilowatts, up to a few hundred kilowatts.

Two parameters that can be useful during a preliminary assessment of different expanders are the specific speed and the specific diameter. These parameters are essentially non-dimensional parameters that describe the speed and diameter of an expander, respectively. The specific speed N_s is defined as:

$$N_s = \frac{\omega \sqrt{\dot{Q}_4}}{\Delta h_s^{3/4}},$$ (5.45)

and the specific diameter D_s is defined as:

$$D_s = \frac{D \Delta h_s^{1/4}}{\sqrt{\dot{Q}_4}},$$ (5.46)

where ω is the rotational speed in rad/s, $\dot{Q}_4$ is the volumetric-flow rate at the expander outlet in m/s^3, Δh_s is the isentropic enthalpy drop across the expander in J/kg (i.e. $\Delta h_s = h_3 - h_{4s}$) and D is the characteristic diameter in m. For example, the characteristic length for a radial turbine is the rotor diameter, whilst the characteristic diameter for a reciprocating-piston expander would be the cylinder diameter.

When N_s and D_s are correlated against expander efficiency, a useful map is obtained that can be used during preliminary expander design and selection. An example of such a chart is provided in Figure 5.18, which relates specifically to a radial turbine. Although it should be noted that more general charts, relevant to a larger number of expanders, are also found within the literature. For example, the study by Badr et al. (1984) provides an example of the use of such maps to investigate expander selection for ORC systems. Referring back to Figure 5.18, this map can ultimately be used to determine the optimal rotational speed N and rotor diameter D for a radial turbine. This is possible since $\dot{Q}_4$ and Δh_s are both known from a thermodynamic optimisation, and N_s and D_s selected from Figure 5.18 give the optimal turbine efficiency. Using N_s and D_s therefore allows the designer to quickly assess the feasibility of using different expander technologies.

5.4.4   Heat Exchangers

After completing a thermodynamic cycle analysis, the next step in the heat-exchanger analysis is to estimate the required heat-transfer areas. To do this it is necessary to

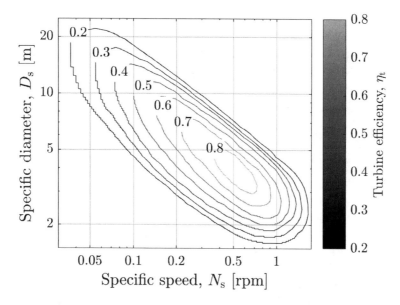

Figure 5.18 Correlation between specific speed N_s and specific diameter D_s against isentropic efficiency for a radial turbine. Reproduced from data extracted from Watson and Janota (1982)

implement a heat-exchanger sizing model. In Section 2.2, we have already introduced the fundamentals of heat-exchanger sizing and selection for waste-heat recovery applications, and the aim of this section is to expand on this analysis in more detail, and apply it specifically to ORC systems.

Heat-Exchanger Sizing Model

Within the evaporator and condenser within a subcritical, non-recuperated ORC system there can be up to five heat-transfer processes. In the evaporator there is single-preheating, two-phase evaporation and finally single-phase superheating if required. Similarly, in the condenser there is two-phase condensation and single-phase precooling if required. Each of these heat-transfer processes can be decoupled from each other and thought of as individual heat exchangers, whose heat-transfer areas can be determined using the analysis presented in Section 2.2 (Equations 2.34 and 2.35). Both the heat exchanger load $\dot{Q}$ and the log-mean temperature-difference $\Delta T_{\log}$ are known from the thermodynamic cycle analysis. Therefore, the heat-transfer areas can be determined if the overall heat-transfer coefficient U for each heat-transfer process is known. This parameter is a function of the properties of the working fluid and the heat source or heat sink, in addition to the geometry of the heat exchanger. As discussed in Section 2.2, the three most common types of heat exchanger are tube-in-tube, shell-and-tube and plate heat exchangers, and these three different types of heat exchanger are typically found within ORC systems. For small-scale applications it is likely that a tube-in-tube or a plate heat exchanger could be used; however, for a larger-scale application a shell-and-tube heat exchanger will probably be more suitable.

Table 5.3 Overall heat-transfer coefficients assumed for the different heat-transfer regions for a shell-and-tube heat exchanger.

Region	Hot Side	Cold Side	U [W/(m^2 K)]
Preheating	Exhaust gas	Working fluid (liquid)	99
Evaporation	Exhaust gas	Working fluid (two-phase)	99
Superheating	Exhaust gas	Working fluid (vapour)	93
Desuperheating	Working fluid (vapour)	Water	105
Condensation	Working fluid (two-phase)	Water	764

Values take from Hewitt (1994)

The simplest way to determine U for different heat-exchanger geometries is to refer to the literature in which attempts have been made to quantify the expected overall heat-transfer coefficients within different heat exchangers operating under different conditions. As an example, estimates taken from Hewitt (1994) for the overall heat-transfer coefficient for different heat-transfer processes within a shell-and-tube heat exchanger are reported in Table 5.3. Using these values it is then simple to determine the required heat-exchanger area.

The alternative to using assumed values for U is to develop a more detailed heat-exchanger model. Depending on the heat exchanger selected, a different heat-exchanger model will be required to estimate U, and this model should in turn be used to optimise the design of the heat exchanger to minimise the heat-transfer area and the pressure drop. Obviously, more the increase in complexity in the heat-exchanger geometry, more complex the heat-exchanger model. For example, a tube-in-tube heat exchanger is relatively simple to model, whilst a shell-and-tube heat exchanger is more advanced since a large number of design variables are introduced.

As with the topic of expander design, it is not the purpose of this book to delve too much into the topic of heat exchanger design and selection. For that purpose, readers should refer to one of the number of textbooks that specifically address this topic (e.g. Kraus (2003) or Verein Deutscher Ingenieure (2010)). Instead, effort will be directed towards developing a simple model for a tube-in-tube heat exchanger to demonstrate the key considerations within the context of a Rankine cycle power system. Examples that demonstrate the integration of other types of heat exchanger within an ORC system include the works of Quoilin et al. (2011), Karellas, Schuster and Leontaritis (2012) and Pierobon et al. (2013).

A schematic of a tube-in-tube of heat exchanger is given in Figure 5.19, in which the notation used to describe it is also defined. Recapping from Section 2.2, for this type of heat exchanger we have:

$$\dot{Q} = \dot{m}_{wf}(h_i - h_o) = \dot{m}_h c_{p,h}(T_{hi} - T_{ho}) = UA\Delta T_{log}, \qquad (5.47)$$

where $\dot{Q}$ is the heat-transfer rate; $\dot{m}_{wf}$ and $\dot{m}_h$ are the working-fluid and heat-source mass-flow rates, respectively; $c_{p,h}$ is the heat-source specific-heat capacity; h_i and h_o

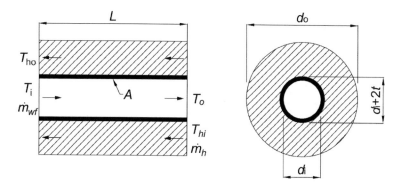

Figure 5.19 Schematic of a tube-in-tube heat exchanger and the designated notation

are the working-fluid inlet and outlet enthalpies, respectively; T_{hi} and T_{ho} are the heat-source temperatures at the inlet and outlet; and $\Delta T_{\log}$ is the log-mean temperature difference, and is given by:

$$\Delta T_{\log} = \frac{\Delta T_1 - \Delta T_2}{\ln\left(\frac{\Delta T_1}{\Delta T_2}\right)}. \tag{5.48}$$

where

$$\Delta T_1 = T_{hi} - T_o; \tag{5.49}$$
$$\Delta T_2 = T_{ho} - T_i. \tag{5.50}$$

As shown in Figure 5.19, a tube-in-tube heat exchanger is constructed from two concentric tubes with inner diameters d_i and d_o, respectively. The two concentric tubes form an internal flow passage through the inner pipe, and a second flow passage through the annual gap between the two pipes. In addition to the pipe diameters the wall thickness separating the two flow passages t is also defined. The ORC working fluid enters the inner pipe at T_i and is heated up to T_o by absorbing heat from the heat source. The heat source travels through the annular space in the opposite direction and is cooled down from T_{hi} to T_{ho}. The required heat-transfer area is defined as:

$$A = \pi d_i L, \tag{5.51}$$

where L is the length of pipe required.

The overall heat-transfer coefficient for a tube-in-tube heat exchanger can be predicted by decomposing the heat-transfer process from the heat source to the working fluid into three separate processes: firstly, there is a convective heat transfer between the heat source flowing through the annular space and the outer surface of the inner pipe; secondly, there is a heat conduction through the inner pipe wall; and thirdly, there is another convective heat-transfer process between the inner wall of the inner pipe and the ORC working fluid. Expressed mathematically, the overall heat-transfer coefficient is then given as:

$$\frac{1}{U} = \frac{1}{\alpha_w} + \frac{A \ln\left(\frac{d_i + 2t}{d_i}\right)}{2\pi k_{wall} L} + \frac{d_i}{(d_i + 2t)} \frac{1}{\alpha_h}, \tag{5.52}$$

where k_{wall} is the thermal conductivity of the wall, and α_w and α_h are the local heat-transfer coefficients of the working fluid and heat-source fluid, respectively.

The local heat-transfer coefficients can be calculated using empirical correlations that are suitable for single-phase heat-transfer and two-phase evaporation, respectively. Although the model has been formulated here for the evaporator, it is equally applicable to the condenser. In this case suitable correlations for two-phase condensation are also required. A short discussion of suitable heat-transfer correlations is provided in the next subsection. It is noted that it is only the calculation of these local heat-transfer coefficients that distinguishes the calculation process for the different heat-transfer regions from each other.

After using suitable correlations to obtain both α_w and α_h, all of the parameters in Equation 5.52 are known and therefore U can be calculated. From this the required heat-exchanger area A immediately follows. Applying this general calculation procedure separately to the preheating, evaporation and superheating processes within an ORC, the total evaporator area is then given as the summation of the preheating area A_{ph}, evaporation area A_{ev} and superheating area A_{sh}:

$$A_h = A_{ph} + A_{ev} + A_{sh}, \tag{5.53}$$

and similarly the condenser area is given as the summation of the precooling area A_{pc} and the condensation area A_{co}:

$$A_c = A_{pc} + A_{co}. \tag{5.54}$$

As will be shown in the following subsection, the heat-transfer correlations are a function of the transport properties of the fluid, namely, the density, thermal conductivity and viscosity. Furthermore, correlations for two-phase heat transfer are a function of the vapour quality. However, within each heat-transfer region these properties do not remain constant. Therefore, to account for these variations each heat-transfer process can be further discretised into a number of smaller heat-transfer processes. For each discrete element 'i' the heat-exchanger sizing model can then be used to determine the area requirement for that element. Therefore, after discretising each heat-transfer process into n elements, the overall evaporator area is:

$$A_h = \sum_{i=1}^{n} A_{ph,i} + \sum_{i=1}^{n} A_{ev,i} + \sum_{i=1}^{n} A_{sh,i}, \tag{5.55}$$

whilst the condenser area is:

$$A_c = \sum_{i=1}^{n} A_{pc,i} + \sum_{i=1}^{n} A_{co,i}. \tag{5.56}$$

Heat-Transfer Coefficient Correlations

The determination of heat-transfer coefficients is of immense interest to many engineering applications, and as such there is a prolific amount of literature on various correlations for different working fluids, operating conditions and heat-exchanger geometries. A thorough analysis of these different correlations is not a focus of this book and the reader is urged to consult various textbooks if this is of further interest (Kraus 2003; Verein Deutscher Ingenieure 2010). Instead, the problem will be formulated using representative correlations for single-phase heat transfer, two-phase evaporation and two-phase condensation to give the reader a flavour of the calculation process and the parameters involved. In general, heat-transfer correlations are used to find the Nusselt number Nu, which is the ratio of convective to conduct heat transfer, and is mathematically expressed as:

$$Nu = \frac{\alpha d_h}{k},$$ (5.57)

where α is the local heat-transfer coefficient, d_h is the hydraulic diameter and k is the thermal conductivity of the fluid. For a tube-in-tube heat exchanger the area and hydraulic diameter depend upon whether the fluid stream is in the inner pipe or the outer annular space. For the inner pipe:

$$d_h = d_i;$$ (5.58)

$$A = \left(\frac{\pi}{4}\right) d_i^2.$$ (5.59)

and for the annular space:

$$d_h = d_o - (d_i + 2t);$$ (5.60)

$$A = \left(\frac{\pi}{4}\right) (d_o^2 - (d_i + 2t)^2).$$ (5.61)

The Gnielinski correlation can be used for single-phase turbulent flow within tubes, and is given as:

$$Nu = \frac{\left(\frac{f}{8}\right)(Re - 1000)Pr}{1 + 1.2\left(\frac{f}{8}\right)^{0.5}(Pr^{\frac{2}{3}} - 1)},$$ (5.62)

where Re is the Reynolds number, Pr is the Prandtl number and f is the Petukhov friction factor. For the correlation to be applicable, $0.5 \leq Pr \leq 2{,}000$ and $3 \times 10^3 < Re < 5 \times 10^6$. The Reynolds number is the ratio of inertial to viscous forces and is given by:

$$Re = \frac{\rho c d_h}{\mu},$$ (5.63)

where ρ and μ are the density and dynamic viscosity of the fluid and c is the fluid velocity. The Prandtl number is the ratio of momentum to thermal diffusivity and is given as:

$$Pr = \frac{c_p \mu}{k},$$ (5.64)

where c_p is the specific-heat capacity of the fluid. All thermophysical properties used within the calculation of these non-dimensional parameters are evaluated at the bulk mean-fluid temperature. Finally, the Petukhov friction factor f is found based on the Reynolds number:

$$f = (0.790 \ln \mathrm{Re} - 1.64)^{-2}. \tag{5.65}$$

Chen's correlation for saturated flow boiling remains one of the most successful correlations for saturated boiling. The ORC working fluid local heat transfer coefficient α_w due to boiling is thought to be the summation of forced convection α_{fc} and nucleate boiling terms α_{nb}:

$$\alpha_w = \alpha_{nb} + \alpha_{fc}. \tag{5.66}$$

The forced convection term is obtained from:

$$\frac{\alpha_{fc} d_h}{k_l} = 0.023 Re^{0.8} Pr_l^{0.4} F, \tag{5.67}$$

where the 'l' subscript refers to Prandtl number based on the saturated-liquid properties, and the vapour quality q is used to scale the saturated-liquid Reynolds number:

$$Re = Re_l(1 - q). \tag{5.68}$$

The term F is given by:

$$F = \begin{cases} 1 & \text{for } \frac{1}{X_{tt}} < 0.1 \\ 2.35 \left(0.213 + \frac{1}{X_{tt}}\right)^{0.736} & \text{for } \frac{1}{X_{tt}} > 0.1, \end{cases} \tag{5.69}$$

where X_{tt} is the Matinelli parameter:

$$X_{tt} = \left(\frac{\rho_v}{\rho_l}\right)^{0.5} \left(\frac{\mu_l}{\mu_v}\right)^{0.1} \left(\frac{1-q}{q}\right)^{0.9}, \tag{5.70}$$

and the subscript 'v' refers to the saturated-vapour properties. The nucleate-boiling term α_{nb} is found by the correlation:

$$\alpha_{nb} = 0.00122 \left[\frac{k_l^{0.79} c_{p,l}^{0.45} \rho_l^{0.49}}{\sigma^{0.5} \mu_l^{0.29} h_{lv}^{0.24} \rho_v^{0.24}}\right] \Delta T^{0.24} \Delta P^{0.75} S, \tag{5.71}$$

where σ is the surface tension, h_{lv} is the latent-heat of vaporisation and S is Chen's suppression factor:

$$S = \left[1 + (2.56 \times 10^{-6})(Re_l F^{1.25})^{1.17}\right]^{-1}. \tag{5.72}$$

The temperature difference ΔT is the difference between the evaporator wall temperature and the bulk-fluid temperature:

$$\Delta T = T_{wall} - T, \tag{5.73}$$

whilst the pressure difference ΔP is the difference between the fluid saturation pressure that corresponds to the wall temperature and the bulk fluid pressure.

$$\Delta P = P_{wall} - P. \tag{5.74}$$

Shah's correlation is one of the most widely used correlations for condensation of a pure saturated vapour in horizontal tubes. It is based on a two-phase multiplier concept in which the two-phase heat transfer of the working fluid α_w is the heat-transfer coefficient for a saturated liquid α_l multiplied by a function of the vapour quality q:

$$\frac{\alpha_w}{\alpha_l} = (1-q)^{0.8} + \frac{3.8q^{0.76}(1-q)^{0.04}}{\left(\frac{P}{P_{cr}}\right)^{0.38}}, \qquad (5.75)$$

where P_{cr} is the fluid critical pressure, and α_l is obtained using the Dittus-Boelter correlation for single-phase flow:

$$\frac{\alpha_l d_h}{k_l} = 0.023 Re_l^{0.8} Pr_l^{0.4}. \qquad (5.76)$$

The subscript 'l' again denotes that saturated liquid properties should be used. For the correlation to be applicable, $10.8 < \rho u < 1{,}599$ kgm^2/s; $Re_l > 350$; $0.02 < P/P_{cr} < 0.44$; $Pr_l > 0.5$; and $0 < q < 1$.

Pressure-Drop Correlations

It should be noted that during the ORC cycle analysis pressure drops are neglected. However, during the optimisation of the heat-transfer area the optimisation will converge on a solution with very small pipe diameters since this is favourable to maximise the heat-transfer coefficient. However, in reality such a design would incur very large pressure drops. Therefore, when optimising the heat-transfer area it is necessary to estimate the pressure drops within each heat-transfer region and then apply constraints to stop the total pressure drops within the evaporator and condenser becoming too large. Expressed mathematically, these constraints are:

$$\Delta P_{ph} + \Delta P_{ev} + \Delta P_{sh} < \Delta P_{max}; \qquad (5.77)$$

$$\Delta P_{pc} + \Delta P_{co} < \Delta P_{max}; \qquad (5.78)$$

$$\Delta P_h < \Delta P_{max}; \qquad (5.79)$$

$$\Delta P_c < \Delta P_{max}, \qquad (5.80)$$

where ΔP_{ph}, ΔP_{ev}, ΔP_{sh}, ΔP_{pc} and ΔP_{co} are the pressure drops within the preheating, evaporation, superheating, precooling and condensation regions, ΔP_h and ΔP_c are the heat-source and heat-sink pressure drops, and ΔP_{max} is the maximum pressure drop constraint.

The pressure drop for a single-phase internal flow of any flow regime through a pipe of length L can be expressed by:

$$\Delta P = f \frac{L}{d_h} \frac{\rho_{av} c^2}{2}, \qquad (5.81)$$

where f is the friction factor (Equation 5.65), d_h is the hydraulic diameter, and $\rho_{av} c^2/2$ is the dynamic pressure and:

$$\rho_{av} = \frac{\rho_{in} + \rho_{out}}{2} \; ; \tag{5.82}$$

$$c = \frac{\dot{m}}{\rho_{av}A} \; . \tag{5.83}$$

This equation is suitable for the calculation of the pressure drop within the preheating, superheating and precooling stages for the working fluid, and for the complete heat-source and heat-sink fluid flows. Within the evaporation and condensation heat transfer stages, the working fluid is two-phase and an alternative pressure drop calculation is required. There are a number of correlations within the literature; however, the Müller-Steinhagen and Heck correlation has been proven to provide suitably accurate results:

$$\frac{\Delta P}{L} = \Lambda(1 - q)^{1/3} + \left(\frac{dP}{dz}\right)_{zo} q^3, \tag{5.84}$$

where

$$\Lambda = \left(\frac{dP}{dz}\right)_{lo} + 2\left[\left(\frac{dP}{dz}\right)_{zo} - \left(\frac{dP}{dz}\right)_{lo}\right] q; \tag{5.85}$$

$$\left(\frac{dP}{dz}\right)_{lo} = f_l \frac{2G^2}{d_h \rho_l} \; ; \tag{5.86}$$

$$\left(\frac{dP}{dz}\right)_{zo} = f_v \frac{2G^2}{d_h \rho_v} \; ; \tag{5.87}$$

$$f = \frac{0.079}{Re^{0.25}} \; ; \tag{5.88}$$

$$Re = \frac{G d_h}{\mu} \; ; \tag{5.89}$$

$$G = \rho c = \frac{\dot{m}}{A} \; , \tag{5.90}$$

and the friction factors for the liquid and vapour phases are found by calculating the Reynolds number using the liquid and vapour viscosity of the fluid, respectively.

5.5   Advanced Optimisation and System Design

Within this chapter we have discussed thermodynamic modelling of the Rankine cycle, working-fluid selection and component selection. To conclude this chapter these various aspects can be brought together and combined, leading to the formulation of an advanced optimisation framework in which optimal systems can be identified for a particular application. The development of such a framework is advantageous to an individual or organisation looking to install a Rankine cycle to generated electricity from a waste-heat stream in that it allows important performance estimations to be obtained.

5.5.1   Optimisation

When considering the design of a Rankine cycle for a particular application, it is necessary to identify the best design that can result in the best performance. Thus, optimisation can be used as a powerful technique to identify the most optimal solutions for a specific case. Of course, the objective of the optimisation may change, depending on whether performance is assessed in terms of thermodynamic performance or an economic performance indicator. For example, one may want to identify the optimal system that maximises power output or thermal efficiency, or alternatively one may be more interested in minimising the specific-investment cost (SIC) or payback. In such optimisations, where there is one objective, the optimisation is referred to as a single-objective optimisation. Alternatively, optimisation can also be used to investigate the trade-off between two objective functions, such as the one between thermodynamic and economic performance, through the use of multi-objective optimisation.

It is not the purpose of this section to explore optimisation methods in detail, but instead to formulate the optimisation problem in relation to a Rankine cycle system. Existing optimisation tools can then be used to conduct the optimisation. Here, this optimisation will be formulated in the form of a multi-objective function that investigates the trade-off between power output and heat-exchanger area, but this formulation is equally valid for other objective functions, and for a single-objective optimisation if the second objective and its related components are simply removed. Thus, the general optimisation is formulated as:

$$\min_{\mathbf{x},\mathbf{y},\mathbf{z}} \{-\dot{W}_n(\mathbf{x},\mathbf{y},\mathbf{z}),\ A(\mathbf{x},\mathbf{y},\mathbf{z})\}, \qquad (5.91)$$

subject to:

$$g(\mathbf{x},\mathbf{y},\mathbf{z}) \leq 0; \qquad (5.92)$$
$$h(\mathbf{x},\mathbf{y},\mathbf{z}) \leq 0; \qquad (5.93)$$
$$k(\mathbf{x},\mathbf{y},\mathbf{z}) \leq 0; \qquad (5.94)$$
$$\mathbf{x}_{min} \leq \mathbf{x} \leq \mathbf{x}_{max}; \qquad (5.95)$$
$$\mathbf{y}_{min} \leq \mathbf{y} \leq \mathbf{y}_{max}; \qquad (5.96)$$
$$\mathbf{z}_{min} \leq \mathbf{z} \leq \mathbf{z}_{max}. \qquad (5.97)$$

In Equations (5.91–5.97), $\mathbf{x}$, $\mathbf{y}$ and $\mathbf{z}$ are the vectors defining the variables that control the thermodynamic cycle, the cycle components and the working fluid, respectively; the latter applies in the general case where the working fluid is included within the optimisation, as we will see towards the end of this chapter. Thus, $\mathbf{x}$ relates to cycle variables such as condensation temperature, pressure ratio and amount of superheat; $\mathbf{y}$ relates to component variables such as heat-exchanger and expander dimensions; and $\mathbf{z}$ are variables that may define the working-fluid structure. The expressions $g(\mathbf{x},\mathbf{y},\mathbf{z})$, $h(\mathbf{x},\mathbf{y},\mathbf{z})$ and $k(\mathbf{x},\mathbf{y},\mathbf{z})$ define the cycle, component and working-fluid constraints, which may control aspects such as minimum and maximum operating pressures and temperatures, heat-exchanger pinch points or pressure drops. Finally, $\mathbf{x}_{min} \leq \mathbf{x} \leq$

x_{max}, $y_{min} \leq y \leq y_{max}$ and $z_{min} \leq z \leq z_{max}$ are the bounds imposed for each group of optimisation variables.

5.5.2   Techno-Economic Optimisation

The analysis presented in Section 5.3 is suitable for identifying optimal cycles that can maximise the power output from a defined heat source and heat sink. This approach may result in the most efficient utilisation of waste heat, but could also correspond to high investment costs and a poor return on investment. Therefore, from the point-of-view of a client, it is better to optimise the cycle based on a techno-economic performance indicators such as the SIC, payback period (PB), net-present value (NPV) or levelised cost of electricity (LCOE). To do this it is necessary to determine the associated investment costs for a particular system. Thankfully, the outputs from implementing the component sizing models described in Section 5.4 are estimates of the component sizes (i.e. the pump and expander power ratings and the required heat-transfer area requirements), which can be translated into estimated invested costs.

At present there are only a limited number of ORC installations worldwide, and unfortunately system cost data are not publicly available. However, cost correlations originating from the chemical industry are commonly used in the literature, in an attempt to obtain cost estimates for ORC systems. A well-established method is the module costing technique introduced by Guthrie (Lemmens 2016), which provides the costs of individual components, based on a specific sizing attribute, and on some cost coefficients. By adding the individual component costs the total ORC unit cost is obtained. The costing method applied within this section uses the cost correlations given by Turton et al. (2009), which takes the form:

$$C_p^0 = F10^{(Z_1 + Z_2 \log(X) + Z_3 \log(X)^2)}, \qquad (5.98)$$

where C_p^0 is the component cost in \$; F is a material factor accounting for the component manufacturing; Z_i is the cost coefficient; and X is the sizing attribute. Both Z_i and X vary depending on the type of the equipment selected for each component. Once the cost of each component is estimated, the Chemical Engineering Plant Cost Index can then be used to convert the cost to today's value.

As a representative example, some example cost correlations are provided in Table 5.4, and these cost correlations will be used within this section to demonstrate a technoeconomic optimisation. However, it is important that the reader identifies the most suitable cost correlations for their application based on the components that are being used. In this instance, the reader can refer to Turton et al. (2009) and Seider, Seader & Lewin (2009), or similar, to identify the correct correlations to implement. More specifically, the cost correlations given in Table 5.4 correspond to a centrifugal pump, a radial turbine and tube-in-tube heat exchangers. The pump correlation is applicable for power ratings between 1 and 300 kW, the expander correlation is applicable for power ratings between 100 and 1500 kW and the heat exchanger correlation is applicable for heat-transfer areas between 1 and 10 m². For the analysis presented in this section, in which the main focus is presenting the methodology to the reader, it is assumed that if

Table 5.4 Cost-correlation coefficients

Component	Attribute (X)	F	Z_1	Z_2	Z_3
Pump	$\dot{W}_p$ (kW)	3.24	3.3892	0.0536	0.1538
Expander	$\dot{W}_e$ (kW)	3.50	2.2474	1.4965	−0.1618
Tube-in-tube heat exchanger	A (m^2)	3.29	3.3444	0.2745	−0.0472

Table 5.5 Assumptions made for the techno-economic optimisation

Pump isentropic efficiency	η_p	0.7	–
Expander isentropic efficiency	η_e	0.8	–
Heat-source temperature	T_{hi}	180	°C
Heat-source mass-flow rate	$\dot{m}_h$	1.0	kg/s
Heat-source specific-heat capacity	$c_{p,h}$	4,200	J/(kg K)
Minimum evaporator pinch point	PP_h	10	°C
Heat-sink temperature	T_{ci}	15	°C
Heat-sink mass-flow rate	$\dot{m}_c$	5	kg/s
Heat-sink specific-heat capacity	$c_{p,c}$	4,200	J/(kg K)
Minimum condenser pinch point	PP_c	5	°C
Maximum heat-exchanger pressure drop	ΔP_{max}	50	kPa

the component sizing attribute lies slightly outside of these ranges, the correlations are still broadly applicable. However, for a particular application being evaluated, it will be necessary that the sizes obtained from the system model are applicable to apply the correlations selected.

With the total investment cost estimated, the SIC, PB, NPV and LCOE can be estimated using Equations 4.18–4.21, which were defined in Chapter 4.

Using these cost correlations it is now possible to couple the thermodynamic and component models together and to conduct a technoeconomic optimisation. For this we reconsider the case study defined in Table 5.2, and also introduce the same heat sink that was considered when evaluated working-fluid mixtures. Therefore, in summary, the case study can be summarised by the values given in Table 5.5.

For the optimisation study six working fluids have been selected which have critical temperatures ranging between 101.2 °C for R134a and 318.6 °C for toluene. For each working fluid the system was optimised multiple times, each time with a different optimisation objective. The first objective maximised the net power output from the system, the second objective minimised the SIC of the system and the third objective maximised the NPV of the system after 20 years. For the calculation of the NPV the assumptions listed in Table 5.6 have been defined.

For the optimisation we consider only a simple, subcritical, superheated, non-recuperated ORC operating with a pure fluid. Therefore, we do not consider alternative cycle architectures or working-fluid mixtures. This system is described by four thermodynamic variables, and four component variables which can be optimised.

Table 5.6 Assumptions for the NPV calculation

Cost of electricity	C_e	0.05	$/kWh
Operation and maintenance cost	$C_{o\&m}$	0.01	$/kWh
Operating hours per annum	n	8,000	hours
Discount rate	r	5	%
Technology lifespan	t_{max}	20	years

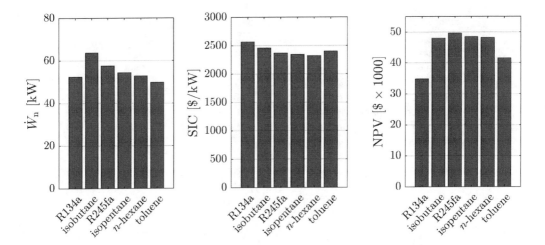

Figure 5.20 Performance of the optimal cycles for six different working fluids when optimised for different objective functions. From left to right: (a) optimal cycles that maximise the net power output $\dot{W}_n$; (b) optimal cycles that minimise the SIC; and (c) optimal cycles that maximise the NPV

The four thermodynamic variables are the condensation temperature T_1, reduced-evaporation pressure P_r, the amount of superheating ΔT_{sh} and the evaporator pinch point at the start of evaporation PP_h, as described previously. The four component variables describe the inner and outer pipe diameters of the evaporator and condenser, respectively, and are defined by the notations $d_{i,h}$, $d_{o,h}$, $d_{i,c}$ and $d_{o,c}$. Alongside the thermodynamic parameters, these four diameters can be optimised to minimise the heat-transfer area, subject to the maximum pressure drop constraint given in Table 5.5.

The results from the optimisations in which the eight variables are optimised for each working fluid, and for each objective function, are summarised in Figure 5.20. The six working fluids are plotted in order of critical temperature, from low to high.

When evaluating Figure 5.20, the main conclusion formulated is that the working fluid that maximises $\dot{W}_n$ is different from the optimal working fluid that minimises the SIC, which in different again to the working fluid that maximises the NPV. More specifically, the maximum power that can be generated by the system is obtained by operating the cycle with isobutane, and this corresponds to net power output of 63.8 kW. The SIC for this cycle is 2483 $/kW. By comparison, selecting n-hexane instead and optimising the cycle to minimise the SIC results in a SIC of 2322 \$/kW, and a power output for 52.7 kW. This corresponds to a 6.5% reduction in the SIC, and a 17.4% reduction in the power output. Clearly, the power output from the system is reduced in the search of minimising the SIC. Finally, the NPV over the 20 year lifespan is maximised by using R245fa, and this corresponds to a NPV of \$49,700. The corresponding values for $\dot{W}_n$ and the SIC are 57.5 kW and 2371 \$/kW, respectively, which are both somewhere between the values for isobutane and n-hexane cycles. In instance, the NPV represents a trade-off between the initial investment cost of the system and the value of the electricity that is generated over the systems lifespan. Ultimately, this analysis shows that when considering economic performance indicators instead of purely thermodynamic ones, alternative cycles and working fluids can be identified, which highlights the importance of considering the components, and associated costs, when evaluating different ORC systems.

5.5.3 Multi-Objective Optimisation

The next step in the optimisation of ORC systems is to move beyond a single objective optimisation in which only $\dot{W}_n$, the SIC or the NPV is optimised, and to consider a multi-objective optimisation (MOO). A MOO is capable of capturing the trade-off between multiple objectives and allows a system designer to make a more informed decision regarding the selection of a particular cycle architecture and working fluid for a particular application. Applied to the design of an ORC system, a MOO allows the trade-off between thermodynamic and economic performance to be investigated. The simplest scenario would be in which one objective is to maximise the $\dot{W}_n$, whilst the second objective is to minimise the total investment cost C_0, although the problem could equally be formulated to optimise a different pair of thermodynamic and economic objectives (i.e. $\dot{W}_n$ and SIC, or thermal efficiency and NPV, etc.).

There are many examples of MOO studies within the literature, with some notable examples being Lecompte et al. (2013); Andreasen et al. (2016); Oyewunmi & Markides (2016). However, most studies are likely to be sensitive to the specific thermodynamic and economic boundary conditions assumed for the study, which makes it hard to formulate general conclusions from the results published within the literature. Instead, it is likely that, for a particular application, a site-specific MOO study should be conducted to identify the most optimal system design. For this purpose, this section will only highlight the purpose of a MOO study within the context of a simple example.

The output from a MOO is a Pareto front, which is shown in Figure 5.21. Within this figure, the results from a MOO optimisation are presented in which the two objectives were to maximise $\dot{W}_n$ and minimise C_0. This optimisation was completed considering the same heat source, heat sink and assumptions that were defined in Section 5.5.2, and considered using isobutane as the working fluid. In Figure 5.21, the Pareto front is described by a series of optimal points, marked by the black circles. The point on the far left represents an optimal cycle that minimises C_0, but also corresponds to the lowest value for $\dot{W}_n$. On the other hand, the point furthest to the right represents an

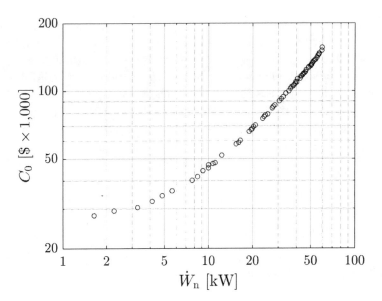

Figure 5.21 Pareto front obtained from a MOO of the net power output $\dot{W}_n$ and total investment cost C_0 for the case study defined in Section 5.5.2

optimal cycle that maximises $\dot{W}_n$, but also corresponds to the highest value for C_0. All of the points in between these two points represent the optimal cycles that correspond to trade-off between the two objectives. For example, using Figure 5.21 it would be possible for a site operator to determine the optimal cycle that would generate the maximum power, based on the amount of investment they have available to invest into a waste-heat recovery technology.

5.5.4   Advanced Working-Fluid Selection

So far the optimisation studies completed within this chapter have considered a predefined working fluid. However, to ensure a truly optimal solution for a given application is obtained, working-fluid selection should be integrated into the optimisation framework. Conventionally, the selection of a working fluid is conducted through a screening study. In such a study a large array of known available working fluids are screened based on pre-determined criteria (e.g. thermodynamic properties, environmental and safety properties, availability), and for each working fluid that passes these criteria a thermodynamic optimisation can be completed. The optimal working fluid is then selected by comparing the results from each optimisation.

An alternative to a conventional working fluid screening study is to introduce computer-aided molecular design (CAMD) into the system model. In a CAMD-ORC model, a group-contribution equation of state is combined with a thermodynamic description of the ORC system. The group-contribution equation of state allows a working fluid to be described by the functional groups from which it is composed (e.g. the working fluid n-pentane can be described by $-CH_3$ and $-CH_2-$ groups). In

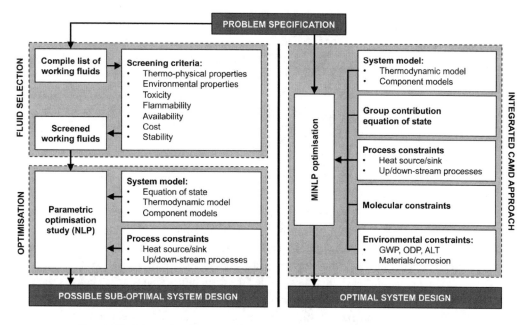

Figure 5.22 Schematic of a conventional ORC optimisation study (left) and the integrated CAMD-ORC approach (right). First reported in White et al. (2017)

this way, the user needs only to define the groups to consider within the optimisation, and integer optimisation. variables can be used to describe the number of each functional group present within the molecule. These integer optimisation variables can then be simultaneously optimised alongside continuous variables describing the thermodynamic system. This facilitates the combined optimisation of the working fluid and the power system, thus removing subjective screening criteria and simultaneously moving towards the next generation of working fluids and optimised ORC systems for waste-heat recovery applications. The key features of the CAMD-ORC problem are compared to a conventional working-fluid selection study in Figure 5.22.

Papadopoulos, Stijepovic and Linke (2010; Papadopoulos et al. 2013) were one of the first to apply CAMD to ORC system design, using it to simultaneously consider aspects such as thermodynamic performance, cost, toxicity and flammability. Since then, the application of CAMD to ORC system design has been gaining traction. The main requirement for conducting CAMD is to have a suitable equation of state that can model a molecule based on its constituent groups. Cubic equations of state, such as the Peng-Robinson (Peng & Robinson 1976) and Redlich-Kwong-Soave (Soave 1972) models, are good choices, and these have been used by a number of researchers (Brignoli & Brown 2015; Cignitti et al. 2017; Su, Zhao & Deng 2017). Another choice is to apply more sophisticated equations of state, based on SAFT (Chapman et al. 1990), which have the advantage of being able to model fluids for which experimental data may not be available. Specifically, CAMD-ORC models have been developed based on the PC-SAFT equation of state (Lampe et al. 2014, 2015; Schilling, Lampe & Bardow 2016; Schilling et al. 2017), and using the SAFT-γ Mie equation of state (White et al. 2017, 2018; van Kleef et al. 2018). The focus of both groups of research has been to integrate heat-exchanger sizing models into the CAMD-ORC model, which allows for economic performance indicators to be introduced, and multi-objective optimisation to be completed in order to identify optimal working fluids and cycle architectures.

5.6   Summary

After having established that a heat engine could represent one of the most promising heat-utilisation technologies for a particular heat source, through preliminary technology-agnostic calculations, it is necessary to conduct a more rigorous assessment of the potential performance of a such a system. To this end, this chapter has presented all of the elements that are required to assess, optimise and select the key components and cycle parameters for a Rankine cycle heat to power system.

In the first instance, this has included the formation of a basic thermodynamic analysis of a simple Rankine cycle, which has introduced the key cycle variables and their effect on cycle operation. Moreover, this analysis has been extended to a number of variants of the Rankine cycle, including recuperated, supercritical and partially evaporated cycles that seek to further improve thermodynamic performance. Alongside this, we have also addressed working-fluid selection, and the role that working-fluid mixtures could have, and provided an introduction to preliminary component design, with a specific emphasis on expander selection, heat-exchanger sizing and component costing. When all of these elements are combined and integrated with a suitable optimisation process, a single optimisation framework can be formed that can be used to identify the optimal fluid, operating conditions and components that can obtain the optimal system for a defined heat source and heat sink. This could be based on thermodynamic performance, economic performance or a trade-off between the two.

6   Heat Pumps and Chillers

6.1   Introduction

While the previous chapter presented a range of heat-to-power technologies that can recover heat over a range of suitable temperatures and convert this into useful work, this chapter provides an overview of existing solutions for providing effective cooling or heating from heat streams originating either from industrial processes (e.g. heat rejected from power stations) or renewable sources (e.g. solar or geothermal sources).

The direct use of heat, for example for industrial process heating or for district heating applications, is the simplest solution for preventing energy waste. The major hurdle here, beyond the identification, proximity and economics of matching heat sources to sinks, is that a significant temperature difference must exist to transfer the thermal energy from the source to the space to be heated, or the sink. As an example, central heating systems operate at temperatures as high as 60–80 °C to heat up spaces commonly around 20 °C. The inability to use low-grade (i.e. low-temperature) heat streams directly is widespread, though it does not make low-quality energy sources useless. Heat pumps and chillers offer viable solutions, both from an energy and economical point of view, to turn low-quality heat into useful outputs, for use in industrial and residential applications.

Heat pumps are a practical solution for upgrading low-grade waste heat to higher temperature levels for use in energy-intensive processes such as drying, distillation processes, district heating or co-generation (i.e. combined heat and power production), whilst absorption chillers can provide efficient cooling, for example for district cooling, or cold storage in the food and pharmaceutical industries. Despite their high potential to help reduce primary energy consumption in heat-intensive applications, heat-upgrading technologies are challenging to design. Different types of heat-pumping systems exist, made of various components and that can be operated with a large number of working fluids. Selecting and sizing suitable systems for particular applications depends strongly on the required heating (or cooling) power and, also importantly, on the temperature level.

Figure 6.1 presents the useful heat consumption in European industrial applications. The annual European industrial heat demand, of about 2,000 TWh in total, is distributed among many sectors (see Figure 6.1a), and is required at various temperatures but mainly (i.e. the largest component of this) from 60 to 150 °C, whilst renewable or waste heat is available within a much larger range, from 30 up to 1,000 °C.

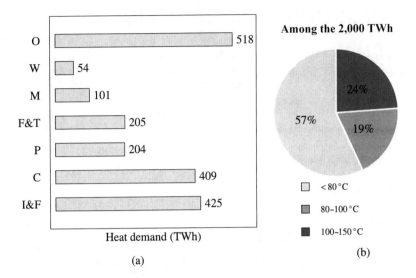

Figure 6.1 Useful heat consumption in European industrial applications: (a) distribution of the yearly thermal energy demand by industrial sector (I&F = Iron and Steel; C = Chemical; P = Paper; F&T = Food and Tobacco; M = Machinery; W = Wood; O = Others) – that sums up to a total of about 2,000 TWh per year in Europe; (b) distribution of the relative heat demand by temperature range. Data taken from Nellissen and Wolf (1983)

Each application presents specific challenges, for both technology innovation, design, development and selection, and for the installation or integration and operation of the resulting technologies into real processes and systems.

This chapter presents heat-pump and chiller technologies that utilise and convert low-grade waste heat, with a focus on: (i) assessing their performance and scope of use; and (ii) selecting adequate designs and working fluids for different applications.

6.2   Working Principle and Performance Indicators

A dish served hot straight from the oven cools down, while ice cream melts when left on the table. In other words, heat transfer through the boundary between two systems at different temperatures occurs spontaneously from the warmer body to the colder one. In this chapter, we examine the working principle and existing designs of devices that move thermal energy in the opposite direction, by absorbing heat from a cold source and rejecting it to a hot sink, through the addition of a driving energy form, thermal energy or work (e.g. mechanical or electrical). Even though generic, the term *heat pump* usually refers to a heating system, the aim of which is to warm up a heat sink, while a *chiller* (or refrigerator) aims to cool down a heat source. If both the heating and cooling capacities are useful (e.g. to provide both swimming-pool water heating and space cooling), the device is sometimes referred to as a thermo-fridge pump.

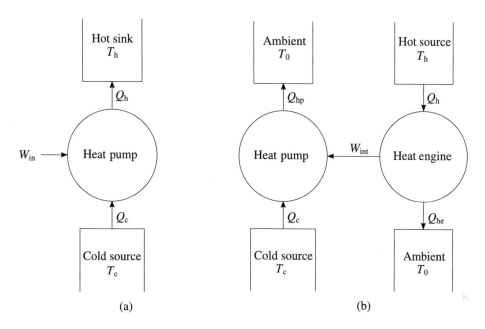

Figure 6.2 Thermodynamic principle of (a) a mechanical or power-driven heat pump, and (b) a heat-actuated heat pump or chiller. All energy, heat and work flow terms are defined positive in this chapter, regardless of the transfer direction

Electrically driven compressors are commonly used to supply the additional external energy in the form of useful electrical work to heat pumps; however, if high-temperature heat is available, the heat can be used to power a heat engine, which in turn can drive a heat-pump system. Figure 6.2 presents schematic diagrams of such mechanical and heat-driven heat pumps or chiller working principles. The heating capacity of the mechanical heat-pump system (see Figure 6.2a), defined as the amount of heat, Q_h, delivered to the hot sink at T_h, ideally (i.e. provided that heat losses are negligible) equals the sum of the heat extracted from the cold source, Q_c, and of the primary energy input, W_{in}:

$$Q_h = W_{in} + Q_c, \tag{6.1}$$

whilst the heat and work flows in a heat-driven system are related according to the following energy balances:

$$Q_c + W_{int} = Q_{hp}; \tag{6.2}$$
$$Q_h = W_{int} + Q_{he}. \tag{6.3}$$

The performance of such devices can be measured by the ratio of the useful energy output (heating or cooling) to the primary energy input. The definition of the so-called coefficient of performance (COP), denoted ϕ for work-driven systems and ψ for

Table 6.1 Coefficient of performance for power- or heat-driven heat pumps and chillers

	Power-Driven	Heat-Driven
Definition	$$\phi_{hp} = \frac{Q_h}{W_{in}}$$ $$\phi_{ch} = \frac{Q_c}{W_{in}}$$	$$\psi_{hp} = \frac{Q_{hp} + Q_{he}}{Q_h}$$ $$\psi_{ch} = \frac{Q_c}{Q_h}$$
Carnot	$$\phi_{hp} = \frac{T_h}{T_h - T_c}$$ $$\phi_{ch} = \frac{T_c}{T_h - T_c}$$	$$\psi_{hp} = \left(1 - \frac{T_c}{T_h}\right)\frac{T_0}{T_0 - T_c}$$ $$\psi_{ch} = \left(1 - \frac{T_0}{T_h}\right)\frac{T_c}{T_0 - T_c}$$
Endo-reversible (Novikov 1958; Velasco et al. 1997)	$$\phi_{hp} = \sqrt{\frac{T_h}{T_h - T_c}}$$ $$\phi_{ch} = \sqrt{\frac{T_h}{T_h - T_c}} - 1$$	$$\psi_{hp} = \sqrt{\frac{T_0}{T_h}}\left(1 - \frac{\sqrt{T_0} - \sqrt{T_h}}{\sqrt{T_0} - \sqrt{T_c}}\right)$$ $$\psi_{ch} = \left(1 - \sqrt{\frac{T_0}{T_h}}\right)\left(\sqrt{\frac{T_0}{T_0 - T_c}} - 1\right)$$

heat-driven ones, differs depending on whether the system is a heat pump or a chiller, although they are thermodynamically equivalent:

$$
\text{COP}
\begin{cases}
\phi_{ch} = \dfrac{Q_c}{W_{in}} \quad \text{and} \quad \psi_{ch} = \dfrac{Q_c}{Q_h} & \text{for a chiller;} \\[2em]
\phi_{hp} = \dfrac{Q_h}{W_{in}} \quad \text{and} \quad \psi_{hp} = \dfrac{Q_{hp} + Q_{he}}{Q_h} & \text{for a heat pump.}
\end{cases}
\tag{6.4}
$$

As the nature of the primary energy input differs whether power- or heat-driven systems are considered, the theoretical values of the COP differ as well. The latter were derived in Chapter 2, providing us with upper bounds and realistic estimates of the performance indicators, as summarised in Table 6.1.

6.3   Thermodynamic Modelling

The design of heat-pump and chiller systems differs, depending not only on the temperature levels at which they operate but also on the form of energy used to drive the heat transport. This section presents a range of heat- and work-actuated systems, as well as recent hybrid systems that maintain high performance over wider application ranges, focusing on closed-cycle heat pumps – as open systems (e.g. mechanical or thermal vapour re-compression cycles) are much less common and available only in a small variety of sizes, as emphasised by Nellissen and Wolf (1983).

6.3.1 Mechanical Heat Pumps

Mechanical heat pumps (also referred to as compression heat pumps) are known for their use in refrigeration applications and are also commonly used for heating (via upgrading). In the context of waste-heat utilisation, electrically- or mechanically driven compressors can be used to extract low-grade heat (at a temperature T_{ci}) from a waste-heat source and turn it into a higher-grade useful process heat (at a temperature T_{hi}).

In this section, we illustrate the working principle and discuss the potential and limitations of mechanical heat-pump technologies.

Vapour-Compression Heat Pump

It is difficult to determine with certainty who first proposed the vapour-compression heat pump (VCHP). A theoretically feasible design of a compression-driven refrigerator was proposed and detailed by Evans (1805), while a working system was built and patented by Perkins (1835). The commercialisation of VCHPs, however, started only in the 1950s and grew rapidly after 1970 due to increasing cost of electricity, as pointed out by Hepbasli and Kalinci (2009). The VCHP is, nowadays, the most widespread and mature refrigeration technology, and common in use in many household, industrial and data-centre cooling applications. Yet, the use of mechanical vapour-compression systems as waste-heat-driven heat pumps for high-temperature process heating is growing (Chua, Chou & Yang 2010): VCHPs are promising systems for reducing primary energy consumption through the utilisation of waste heat.

The VCHP working principle is based on the increase of the liquid-vapour saturation temperature with increasing pressure. In a VCHP system (see Figure 6.3), heat is extracted from a cold source and rejected to a hot sink using, respectively, the latent heat absorbed during evaporation and that released during condensation of a working fluid at different temperatures (and pressures). Due to the saturation temperature dependency upon pressure and the temperature difference between T_{ci} and T_{hi}, these phase-change processes occur at different pressures. VCHP systems use a compressor to circulate and pressurise the working fluid from the evaporating pressure, P_4, to the condensing pressure, P_3.

The four main components of a conventional single-stage MVC system are the compressor, condenser and evaporator, and an expansion valve (also referred to as metering device), as shown in Figure 6.3. The thermodynamic cycle undergone by the working fluid in this closed-loop system is a sequence of four thermodynamic processes.

Vapour compression takes place in the compressor (from States 1 to 2), increasing the pressure of the saturated or superheated vapour leaving the evaporator, while reducing its specific volume.

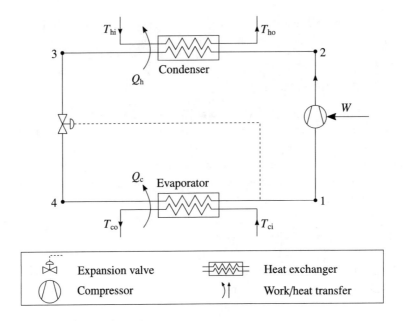

Figure 6.3 Operating principle of a vapour compression cycle

Condensation in the hot-side heat exchanger (States 2 to 3) is ideally an isobaric process, whereby the high-pressure superheated vapour is condensed and potentially subcooled.

Liquid throttling through the metering device (States 3 to 4) is an isenthalpic process without work production or heat transfer. The expansion valve is a key component for a sound and highly efficient operation of a VCHP installation. The role of this metering device is to control the rate at which the fluid is dispensed to the evaporator to maintain a preset constant amount of superheat in the suction line of the compressor.

Evaporation in the cold-side heat exchanger (States 4 to 1) is ideally an isobaric process whereby the low-pressure two-phase mixture is evaporated and preferably superheated to avoid the presence liquid in the compressor suction line.

To further describe the operation of a VCHP system, the underlying processes which describe changes in the state of the working fluid throughout the corresponding thermodynamic system can be reported in thermodynamic diagrams. Figure 6.4 presents ideal temperature–entropy (T–s) and pressure–enthalpy (P–h) diagrams of a compressor-driven heat-pump system operating in the conditions stated in Table 6.2. The T–s diagram is convenient to report the temperature change in the cooling and heat-releasing circuits but is not the most adapted visualisation tool for the VCHP cycle. The P–h diagram is widely used to describe the operation of heat-pump cycles, as isobaric heat-addition and heat-removal and isenthalpic throttling processes are easily identified. Plus, it clearly shows the ratio between the heating, Q_h, and cooling capacities, Q_c, which are functions of: (i) the enthalpy difference across the

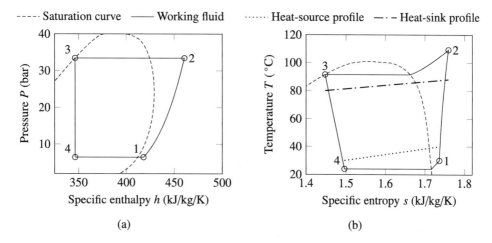

Figure 6.4 Temperature–entropy (T–s) and pressure–enthalpy (P–h) diagrams of a VCHP heat-upgrading cycle using R134a as a working fluid, extracting heat from a low-temperature ($T_{ci} = 40\,°C$) waste-heat source and providing high-temperature ($T_{hi} = 80\,°C$) process heat under the conditions stated in Table 6.2

condensing unit (i.e. $h_2 - h_3$), and (ii) the enthalpy difference across the evaporator (i.e. $h_1 - h_4$), respectively:

$$Q_h = h_2 - h_3\,,\text{ and} \tag{6.5}$$

$$Q_c = h_1 - h_4. \tag{6.6}$$

Common thermodynamic analyses of the VCHP cycle, including that from which the diagrams drawn in Figure 6.4 have been derived, are based on several widely accepted assumptions:

- Pressure losses between the different components are neglected.
- Heat losses to the surrounding are neglected (e.g. overall adiabatic compression).
- The cycle is assumed to operate under steady conditions.
- The working fluid at the exhaust of the condenser is assumed to be saturated liquid.
- Counter-current heat exchangers are used.
- No phase change is permitted in the heat-source and heat-sink streams.

Furthermore, in this type of modelling a set of component performance indicators are used to define the performance of each system component:

- The compressor isentropic efficiency, η_c.
- The evaporator pinch-point temperature difference, PP_c.
- The condenser pinch-point temperature difference, PP_h.

In addition to defining the heat-pump architecture and the working fluid, the VCHP cycle can be fully defined by three thermodynamic parameters. Many sets of independent parameters can be chosen. In this chapter, the VCHP cycle is defined by:

Table 6.2 Working conditions for a compressor-driven heat-pump system

Description	Value
Working fluid	R134a
Heat-source temperature	$T_{ci} = 40\,°C$
Heat-sink temperature	$T_{hi} = 80\,°C$
Pinch-point temperature difference in condenser	$PP_c = 6\,°C$
Pinch-point temperature difference in evaporator	$PP_h = 6\,°C$
Condensation pressure	$P_3 = 33.4$ bar
Evaporation pressure	$P_4 = 6.5$ bar
Superheating at the evaporator exhaust	$\Delta T_{sh} = 6\,°C$

- the condensation pressure, P_3;
- the evaporation pressure, P_4;
- the suction superheat, ΔT_{sh}.

Our analysis starts at the condenser outlet, as this point (State 3) is fully defined by the assumptions and parameters defined above. Indeed, no subcooling is assumed at the exhaust of the condenser (i.e. the vapour quality, q_3, is null), and the condensation pressure, P_3, is known. The thermodynamic State 3 is thus obtained from the working fluid equation of state (EoS):

$$[T_3, h_3, s_3] = \text{EoS} \ (P_3, q_3 = 0, \text{fluid}). \tag{6.7}$$

As the liquid throttling through the metering device is an isenthalpic process ($h_4 = h_3$), the two-phase mixture resulting from the Joule–Thompson expansion undergone from State 3 to 4 can be calculated from the EoS:

$$[T_4, s_4] = \text{EoS} \ (P_4, h_3, \text{fluid}). \tag{6.8}$$

The compressor suction pressure, P_1, equals that at the inlet of the evaporator, P_4, as the heat removal process is assumed adiabatic. And, as the suction superheat, ΔT_{sh}, is known, the evaporator exhaust temperature and state are determined by:

$$T_1 = T_4 + \Delta T_{sh}; \tag{6.9}$$

$$h_1 = \text{EoS} \ (T_1, P_1, \text{fluid}). \tag{6.10}$$

Due to thermodynamic irreversibilities arising from various loss mechanisms (e.g. gas-to-wall in-cylinder heat transfer, pressure losses through the intake and exhaust valves, mass leakage through the piston ring), the working-fluid compression process is not isentropic. This behaviour can be captured by the compressor isentropic efficiency, η_c, that is the ratio of the isentropic to the actual compression work:

$$\eta_c = \frac{W_{c,s}}{W_c}. \tag{6.11}$$

For an overall adiabatic compression (which implies that $W_c = h_2 - h_1$), this ratio can be defined as a function of the discharge-to-suction enthalpy difference:

$$\eta_c = \frac{h_{2s} - h_1}{h_2 - h_1}, \tag{6.12}$$

where h_{2s} is the fictive enthalpy of the working fluid at the discharge of the compressor following an isentropic compression ($s_2 = s_1$) obtained from the known condensation (i.e. discharge) pressure, $P_2 = P_3$:

$$h_{2s} = \text{EoS}\,(P_3, s_1, \text{fluid})\,. \tag{6.13}$$

The actual discharge enthalpy thus follows:

$$h_2 = h_1 + \frac{h_{2s} - h_1}{\eta_c}. \tag{6.14}$$

At that point, all four thermodynamic states defining the VCHP cycle have been obtained. As a mechanically driven machine, the coefficient of performance, ϕ, of compression heat pumps is expressed as a function of the heating capacity and work input:

$$\phi_{\text{VCHP}} = \frac{Q_h}{E_{\text{in}}} = \frac{Q_h}{W_c} = \frac{h_2 - h_3}{h_2 - h_1}. \tag{6.15}$$

To determine the compressor power consumption and the delivered heating power, while utilising a given amount of heat, $\dot{Q}_c$, taken from a fully defined waste-heat stream (i.e. defined inlet temperature, T_{ci}, specific-heat capacity, $c_{p,c}$, and mass flowrate, $\dot{m}_c$) using a VCHP system, the working-fluid mass flowrate, $\dot{m}_{wf}$, must be determined from an energy balance applied to the evaporator:

$$\dot{Q}_c = \dot{m}_c\, c_{p,c}\, (T_{ci} - T_{co}) = \dot{m}_{wf}\, (h_1 - h_4). \tag{6.16}$$

Both the heat-source outlet temperature, T_{co}, and the working-fluid mass flowrate are determined from the equation above, which, in turn, allows to calculate the power consumption, $\dot{W}_c$, and the delivered heating power, $\dot{Q}_h$:

$$\dot{W}_c = \dot{m}_{wf}\, (h_2 - h_1)\,; \tag{6.17}$$

$$\dot{Q}_h = \dot{m}_{wf}\, (h_2 - h_3) = \dot{m}_h\, c_{p,h}\, (T_{ho} - T_{hi})\,. \tag{6.18}$$

The hot heat-sink temperature outlet, T_{ho}, is finally determined using the latter equation.

The final necessary step in the cycle analysis is to check that the temperature profiles in both the heat exchangers do not violate the pinch-point constraints. Denoting by $T_{2\text{-}3}(x)$ and $T_{4\text{-}1}(x)$ the working-fluid temperature profiles in the condenser and evaporator, respectively, and by $T_h(x)$ and $T_c(x)$ the heat-sink and heat-source temperature profiles, respectively, those constraints can be written as:

$$T_{2\text{-}3}(x) - T_h(x) \geqslant PP_h \quad \forall x, \text{ and} \tag{6.19}$$

$$T_{4\text{-}1}(x) - T_c(x) \geqslant PP_c \quad \forall x, \tag{6.20}$$

where x denotes a position in the heat exchanger.

Single-stage VCHP systems are simple, easy to operate and suitable for use in a large number of applications. However, they are limited to relatively low source-to-sink temperature lifts. Large source-to-sink temperature lifts entail large pressure ratios across the compressor, which introduces several technical challenges.

First, a working fluid must be found which can operate over the required range of temperatures and pressures without exceeding critical conditions. If no such fluid exists, one solution is to use a cascade heat-pump system that employs two or more individual compression cycles, operating at different temperature levels with different working fluids, although these systems are more complex and expensive than single-stage systems. Cascade heat pumps are presented in Section 6.3.1.

Provided that a suitable working fluid is found, a second issue is the large pressure ratio required. Valveless positive-displacement (or, volumetric) compressors, such as screw or scroll machines, are typically designed with built-in pressure ratios up to 4 or 5. With these machines, multi-stage compression must be used to provide the required discharge pressure. Reciprocating-piston compressors can be used to achieve higher pressure ratios (up to 20 or 30) but their performance decreases significantly when increasing the pressure lift, which is due to long compression and expansion strokes that leave very short time for the suction and discharge processes to take place. Multi-stage piston compressors are thus also recommended for large pressure ratios, which as earlier, adds to the complexity and cost of the overall system. The performance of volumetric compressors is discussed in further detail in Section 6.6.

Finally, the large discharge temperature attained at the exhaust of the compressor is also a major concern in high-temperature heat pumps, not only as it limits the choice of adequate working fluids, but also because large temperatures induce lubricating oil thermal degradation. Oil is a key element of compression heat pumps, as it (i) ensures lubrication of the moving parts; (ii) helps removing heat generated by the solid friction; and (iii) reduces leakage both between the compression stages in rotating compressors and between the crankcase and the compression chamber in reciprocating-piston machines. An efficient solution to decrease the compressor discharge temperature is to install a flash-tank vapour-injection (FTVI) control loop with a 2-stage compressor.

Flash-Tank Vapour-Injection Heat Pumps

The commercialisation of FTVI systems started in the late 1970s. Since then, the use of this single-fluid two-stage heat pump has markedly increased in high ambient temperature cooling applications. In the context of waste-heat recovery, FTVI heat pumps offer the opportunity to upgrade low-temperature heat sources with a large temperature lift while operating under lower discharge temperature compared to the conventional mechanical vapour-compression cycle.

A vapour-injection heat pump system (presented in Figure 6.5) is a variant of the VCHP that involves the use of a flash tank that splits the cycle into two flow loops, as shown in Figure 6.5, but also as illustrated in the FTVI P–h diagram plotted in Figure 6.6. The high-pressure superheated vapour discharged by the compressor (State 2) is condensed and potentially subcooled in the condenser (State 3) before flowing

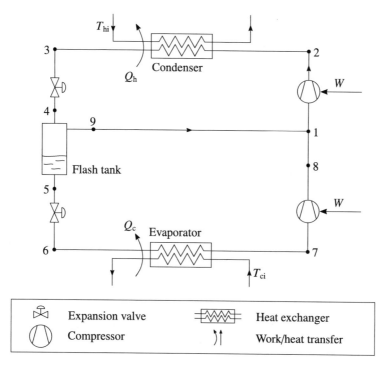

Figure 6.5 Operating principle of a vapour-injection heat pump cycle

through the upper-stage expansion valve, where the fluid undergoes an isenthalpic flash evaporation caused by the sudden drop in pressure, from the condensing to the intermediate pressure. The resulting vapour–liquid mixture (State 4) separates inside the flash tank. The saturated liquid (State 5) is throttled again through the lower-stage expansion valve before absorbing heat in the evaporator. The superheated vapour (State 7) is compressed up to the intermediate pressure (State 8) in the first-stage compressor and is mixed with the saturated vapour (State 9) flowing from the flash tank, thus lowering the superheat in the suction line of the second-stage compressor.

The advantages of the FTVI heat pump over the conventional VCHP are numerous. First, as outlined earlier, lower compression discharge temperatures permit their use in a wider range of applications, notably for providing high-temperature process heat from low heat-source temperatures, which is crucial in the field of waste-heat recovery. Both heat-pump performance and capacity are also increased by using the vapour-injection technique, especially in high-temperature-lift cases. Finally, their heating capacity can be adjusted by controlling the intermediate pressure level, that is, by controlling the opening of the upper-stage expansion valve. This is the reason why, in most cases, the latter is an electronic expansion valve that allows advanced control.

Along with the use of multiple-stage or split compressors, this increased control capability is unfortunately also the main drawback of FTVI systems, which are more complicated to operate, as the sound operation of these two-stage systems requires an adequate liquid level in the flash tank to be maintained.

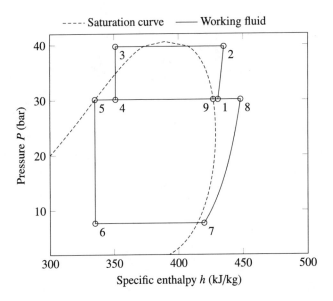

Figure 6.6 Ideal pressure–enthalpy (P–h) diagram of a VIHP heat-upgrading cycle using R134a as a working fluid, extracting heat from a low-temperature ($T_{ci} = 30\,°C$) waste-heat source and providing high-temperature ($T_{hi} = 80\,°C$) process heat under the conditions stated in Table 6.2

Cascade Heat-Pump Systems

Cascade systems are made of two (or more) heat-pump cycles, which use different working fluids with different boiling/condensation points. The heat extracted in a lower-temperature cycle is upgraded up to an intermediate temperature and transferred to a higher-temperature cycle through an intermediate heat exchanger. Energy savings arise from the use of suitable refrigerants with temperature characteristics specifically chosen for each of the lower- and higher-temperature sides. Cascade systems are suitable for large temperature lifts, which arise either in ultra-low-grade waste-heat upgrading applications and/or applications requiring high-temperature heat delivery.

6.3.2   Heat-Actuated Heat Pumps and Chillers

While mechanical work is used as the prime energy source in compression heat pumps, heat-powered refrigeration technologies can provide combined heating and cooling from the thermal energy extracted from a heat source. Several thermally actuated cooling technologies exist (e.g. desiccant cooling), but this section focuses on sorption systems, which are a variety of closed-cycle thermo-chemical heat pump.

The underpinning technological principles, design, operation and characteristics of heat-actuated heat pumps differs largely from those of work-driven heat pumps. The former rely on reversible chemical reactions between media or substance, of a refrigerant (absorbent or adsorbent) onto an absorbate or adsorbate, liquid or solid.

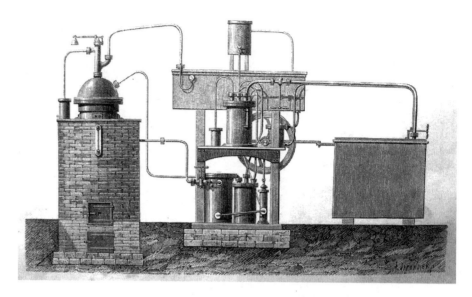

Figure 6.7 Ferdinand Carré's *aqua ammonia* ice-making device. Illustration (in the public domain) retrieved from https://en.wikipedia.org/wiki/Ferdinand_Carre

Absorption Heat Pumps and Chillers

The first gas-absorption refrigeration system using gaseous ammonia dissolved in water (i.e. 'aqua ammonia') was developed and patented by Ferdinand Carré in 1860.

Based on the combined use of a refrigerant and an absorption medium (e.g. ammonia and water), an absorption heat pump or chiller consists of two loops, as shown in Figure 6.8. It is made of five main components, namely: four heat exchangers and a pump. The vapour refrigerant leaving the evaporator (State 10) is absorbed into a solution relatively poor in refrigerant (State 6), while rejecting heat, to form a rich liquid solution (State 1) that is circulated by the pump. In the desorber, part of the dissolved refrigerant (State 7) is evaporated by the heat extracted from a high-temperature heat source, at T_{hi}, and then flows through the condenser (where heat is rejected at a reference temperature T_0) and evaporator (removing heat at a low-temperature T_{ci}), before being absorbed back into the solution. The remaining solution (State 4) is driven through an expansion valve by the pressure difference between the desorber and absorber. An additional heat exchanger can be placed between the cold rich solution leaving the pump (States 2–3) and the hot poor solution leaving the desorber (States 4–5).

Heat is thus removed from two heat sources in the evaporator and desorber, while being rejected, usually at the same reference temperature T_0, in the absorber and condenser. In the context of waste-heat utilisation, absorption systems can be used both as chillers or heat pumps.

An absorption chiller offers the potential to provide low-temperature cooling (Q_{ev}) from harnessing the thermal energy of a high-temperature waste-heat stream (Q_{des}). In situations where both heating and cooling are required, this technology proves

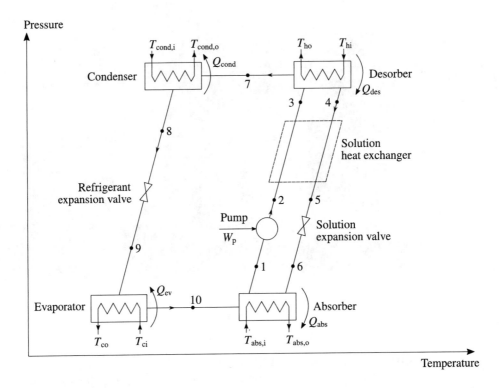

Figure 6.8 Operating principle of an absorption heat pump or chiller

very useful as intermediate-temperature heat can be recovered from the condenser and absorber. The performance analysis of combined heat-cooling absorption systems is, however, not performed here.

An absorption heat pump uses a high-temperature heat source to upgrade low-temperature heat – collected in the evaporator at T_{ci} – into an intermediate-grade heat rejected in the condenser and absorber units at T_0.

An absorption heat pump (or chiller) can be thought of as the superposition of a heat engine, operating between the high-temperature heat source T_{hi} and the reference temperature T_0, and a heat pump (or chiller) operating between T_0 and the cooling temperature T_{ci}. The heat engine, which corresponds to the absorption-medium loop (States 1–6), generates work that drives the heat pump, which corresponds to the refrigerant loop (States 7–10).

The thermodynamic analysis of the absorption chiller cycle considered here, much like all chemical heat-pump cycles, is often based on the following assumptions:

- Pressure losses between the different components are neglected.
- Heat losses to the surrounding are neglected.
- The cycle is assumed to operate under steady conditions.

It is also based on the following indicators that define the performance of each system component:

- the pump isentropic efficiency, η_p;
- the solution heat-exchanger effectiveness, ε;
- the pinch-point temperature difference in each heat exchanger, PP_i, with $i =$ (cond, ev, abs, des).

As shown schematically in Figure 6.8, absorption heat pumps operate between two pressure levels, namely: (i) the high-pressure level, P_{high}, at which occur both the isobaric heat addition process in the desorber and the heat rejection in the condenser; and (ii) the low-pressure level, P_{low}, at which occur both the heat removal in the evaporator and heat rejection in the absorber. From a mathematical point of view, this can be written as:

$$P_1 = P_6 = P_9 = P_{10} = P_{low}; \tag{6.21}$$
$$P_2 = P_3 = P_4 = P_5 = P_7 = P_8 = P_{high}. \tag{6.22}$$

To fully determine the thermodynamic states of both the solution and the refrigerant in the different locations/states throughout the heat pump system, the set of governing equations summarised in Table 6.3 is solved. The explicit solution of this non-linear system of equations is not described here, as absorption heat-pump modelling is often performed using non-causal solvers such as those used in EES or Modelica. The great flexibility offered by non-causal algorithms is very useful due to the large number of equations, as they allow to switch between different sets of input parameters.

Adsorption Heat Pumps and Chillers

The working principle of an adsorption heat pump or chiller is similar to that of absorption systems, except that it relies on solid sorption rather than liquid sorption. A pump is thus not required but valves need to be actuated periodically to guide the different periods of the cycle. The operation and characteristics of adsorption systems are well described in the open-access chapter by Elsheniti et al. (2017).

6.4   Working-Fluid Selection

6.4.1   Mechanical Heat Pumps

Similar to the challenge of working-fluid selection for organic Rankine cycle (ORC) engines, which share many common fluids, the choice of refrigerants for compressor-driven heat pumps is not straightforward, and often multiple candidates are suitable for a specific application. Beyond solely performance-oriented considerations, availability, safety and cost have long guided the choice of refrigeration and air-conditioning refrigerants. In the early 1920s, chlorofluorocarbons (CFC) and hydrochlorofluorocarbons (HCFC), notably used in the Freon trademark products, were the first non-toxic (unlike ammonia) and non-flammable fluids to be used. R-22 was widely used for

Table 6.3 Governing equations for the single-effect absorption heat-pump or chiller

Components	Total Mass Conservation	Refrigerant Conservation	Energy Conservation	Additional Equation
Absorber	$\dot{m}_1 = \dot{m}_{10} + \dot{m}_6$	$\dot{m}_1 x_1 = \dot{m}_{10} x_{10} + \dot{m}_6 x_6$	$\dot{m}_1 h_1 + \dot{Q}_{abs} = \dot{m}_{10} h_{10} + \dot{m}_6 h_6$ $\dot{Q}_{abs} = \dot{m}_{abs}\, c_{p,0}\, (T_{abs,o} - T_{abs,i})$	$T_{6\text{-}1}(x) - T_{abs}(x) \geq PP_{abs}$ $\forall x$ in the absorber
Pump	$\dot{m}_2 = \dot{m}_1$	$x_2 = x_1$	$\dot{W}_p = \dot{m}_1 (h_2 - h_1)$	$\eta_p = \dfrac{h_{2s} - h_1}{h_2 - h_1}$ $h_{2s} = \mathrm{EoS}(P_2, S_1, x_2)$
Desorber	$\dot{m}_3 = \dot{m}_7 + \dot{m}_4$	$\dot{m}_3 x_3 = \dot{m}_7 x_7 + \dot{m}_4 x_4$	$\dot{m}_3 h_3 + \dot{Q}_{des} = \dot{m}_7 h_7 + \dot{m}_4 h_4$ $\dot{Q}_{des} = \dot{m}_{des}\, c_{p,h}\, (T_{hi} - T_{ho})$	$T_{3\text{-}4}(x) - T_h(x) \geq PP_{des}$ $\forall x$ in the desorber
Solution valve	$\dot{m}_6 = \dot{m}_5$	$x_6 = x_5$	$h_6 = h_5$	—
Solution exchanger	$\dot{m}_3 = \dot{m}_2$ & $\dot{m}_5 = \dot{m}_4$	$x_3 = x_2$ & $x_5 = x_4$	$\dot{m}_2 h_2 + \dot{m}_4 h_4 = \dot{m}_3 h_3 + \dot{m}_5 h_5$	$\dot{Q}_{des} = \varepsilon \dot{C}_{min} (T_4 - T_2)$ $\dot{C}_{min} = \min(\dot{C}_{4,5}, \dot{C}_{2,3})$ $\dot{C}_{ij} = \dot{m}_i \dfrac{h_i - h_j}{T_i - T_j}$
Condenser	$\dot{m}_8 = \dot{m}_7$	$x_8 = x_7$	$\dot{Q}_{cond} = \dot{m}_7 (h_7 - h_8)$ $\dot{Q}_{cond} = \dot{m}_{cond}\, c_{p,0}\, (T_{cond,o} - T_{cond,i})$	$T_{7\text{-}8}(x) - T_{cond}(x) \geq PP_{cond}$ $\forall x$ in the condenser
Refrigerant valve	$\dot{m}_9 = \dot{m}_8$	$x_9 = x_8$	$h_9 = h_8$	—
Evaporator	$\dot{m}_{10} = \dot{m}_9$	$x_{10} = x_9$	$\dot{Q}_{ev} = \dot{m}_9 (h_{10} - h_9)$ $\dot{Q}_{ev} = \dot{m}_c\, c_{p,c}\, (T_{ci} - T_{co})$	$T_{9\text{-}10}(x) - T_c(x) \geq PP_{ev}$ $\forall x$ in the evaporator

refrigeration applications, while the R-114 was preferred for high-temperature heat-pumping systems. Due to their large ozone-depletion potential (ODP) and the related danger of increasing the intensity of ultraviolet radiation, the detrimental effect on the environment of CFCs and HCFCs was clearly identified in the early 1970s. The Montreal Protocol, agreed in 1987, is an international agreement that established the timeline for the worldwide ozone-depleting refrigerants, phase out.

Alternative refrigerants that met the requirements of the Montreal Protocol were then favoured, including synthetic, for example hydrofluorocarbons (HFCs), and natural refrigerants, for example hydrocarbons (HCs), ammonia and CO_2 either in sub- or supercritical cycles. HFCs present similar advantages to the CFCs and HCFCs as they are non-toxic and non-flammable, and are classified as zero-OPD working fluids. However, they exhibit high global warming potential (GWP). The Kigali Amendment to the Montreal Protocol, agreed in 2016, entered into force in January 2019 to phase down and, ultimately, phase out the use of these man-made harmful compounds, currently still widely used.

The transition towards zero-OPD and low-GWP working fluids suitable for use in compression-driven heat pumps, such as hydrofluoroolefins (HFOs), ammonia or HCs, however, faces other challenges, related to safety issues and cost. HCs and HFOs are non-toxic but highly flammable fluids, which require specific and expensive equipment and safe practices. The toxicity of ammonia is of great concern when used in partially hermetic systems and in poorly ventilated areas. For these reasons, even though being phased out, HFCs are still considered a valid option for the design of mechanical heat pumps and refrigeration systems. A list of refrigerants commonly used in compression heat pumps is provided in Table 6.4.

Beyond the environmental properties of the fluid, the desired properties of the ideal heat-pump working fluid are similar to those discussed for the ideal ORC systems listed in Chapter 5. The thermodynamic properties of the working fluids must be such that:

- the coefficient of performance, ϕ, is optimal;
- the operating pressure ratio is not excessive, so as to avoid high discharge temperature, which can notably result in oil degradation;
- the evaporation pressure is above atmospheric pressure to avoid operating the condenser under a vacuum;
- the condensation pressure is not excessive to avoid high mechanical stress and expensive component design;
- the minimum ambient temperature is above the fluid triple point;
- high heat-transfer rates are achieved within heat exchangers with minimum temperature differences (high latent heat of evaporation and thermal conductivity);
- pressure drops within the system are low;
- high power density is achieved, notably by using high-density fluids.

In addition to the preferred thermodynamic properties listed earlier, and to cost issue, it is required that the selected working fluid is: (i) non-corrosive and compatible

Table 6.4 Common refrigerants used in compression heat pumps

Fluid	Chemical Formula	GWP	Molecular Weight [g/mol]	Critical Temperature [°C]	Critical Pressure [bar]
Hydrofluorocarbons (HFCs)					
R134a	CH_2FCF_3	1,430	102.0	101.0	38.7
R152a	CH_3CHF_2	124	66.1	113.0	44.5
R410a	$CH_2F_2+C_2HF_5$	2,088	72.6	71.3	48.5
Hydrofluoroolefins (HFOs)					
R1234yf	$CF_3CF=CH_2$	4	114.0	94.7	33.0
R1234ze	$CF_3CF=CHF$	7	114	109.4	35.5
Hydrocarbons (HCs)					
Propane	C_3H_8	3	44.1	97.0	41.35
Butane	C_4H_{10}	4	58.1	150.8	35.9
Cyclopentane	C_5H_{10}	<0.1	70.1	238.5	45.15
Pentane	C_5H_{12}	20	72.1	196.6	33.27
Toluene	C_7H_8	<0.1	92.1	318.6	41.26
Natural refrigerants					
Ammonia	NH_3	0	17.03	133.0	107.43
Carbon dioxide	CO_2	1	44.0	70.0	175.17

with all system materials; and (ii) chemically stable within the system's operating range.

6.4.2 Heat-Actuated Heat Pumps and Chillers

Absorption Heat Pumps and Chillers

Two main working-fluid pairs employed in absorption chillers are: (i) water/lithium bromide (LiBr); and (ii) ammonia/water. The application ranges of water/LiBr and ammonia/water technologies differ significantly, due to the fluids, thermodynamic properties. A well-documented review and analysis of both systems is provided by Herold, Radermacher and Klein (2016). The main observations and comments related to the use of H_2O/LiBr and NH_3O/H_2O as refrigerant/absorption medium pairs is summarised in this section.

Water/Lithium Bromide Absorption Chillers

A H_2O/LiBr absorption chiller uses the LiBr as the absorption medium, while the water plays the role of refrigerant, which limits the cooling temperature above 0 °C. Typically operating with a high-temperature heat source at 70–200 °C, low-cost aqueous LiBr single-effect absorption chillers are commonly used for air-conditioning applications, with coefficients of performance ranging from 0.6 to 1.2.

However, the utilisation of water as a refrigerant causes important technical constraints. For relatively low cooling temperatures, part of the system operated under sub-atmospheric conditions, due to the low water vapour pressure. Typically, for an evaporation temperature of $10\,^{\circ}C$, the chiller operates with a low pressure level of 0.01 bar. This significant vacuum, yet not extreme, is associated with design and operational constraints. Air leaks within the engine are not only detrimental to the overall system performance but also causes corrosion, notably if carbon steel or copper are used.

Ammonia/Water Absorption Chillers

Unlike its $H_2O/LiBr$-based counterpart, the NH_3O/H_2O absorption chiller uses water as the absorption medium, while ammonia serves as a refrigerant. Ammonia/water systems are thus able to operate at much lower temperatures than $H_2O/LiBr$ systems, and can therefore be used for refrigeration, freezing and gas-fired air-conditioning, but also space heating applications, while the latter is beyond the scope of waste-heat recovery. One important drawback of this kind of absorption chiller arises from the toxicity of ammonia, which restricts its use in well-ventilated and non-confined areas.

As shown by Horuz (1998), ammonia/water absorption chillers exhibit COPs that are typically lower than $H_2O/LiBr$ systems. With COPs typically ranging from 0.5 to 0.8, the main advantage of the former working-fluid pair lies in its increased flexibility and potential for waste-heat utilisation. The performance of a single-effect NH_3O/H_2O absorption chiller is discussed in the following section.

Adsorption Heat Pumps and Chillers

Material pairs used in adsorption systems greatly affect the overall system performance. As discussed by Elsheniti et al. (2017), preferred properties for the absorbent are: (i) a wide adsorption capacity range; (ii) high mass and thermal diffusivity; (iii) thermal stability over the operating range; and (iv) low susceptibility to contamination.

Common adsorbent–refrigerant pairs in solid-sorption heat pumps include silica gel–water, zeolite–water and active carbon–ammonia.

6.5 Performance of Heat Pumps and Chillers

The performance of compressor-driven heat pumps and heat-driven chillers is now investigated, using the thermodynamic models presented in Section 6.3.

6.5.1 Vapour-Compression Heat Pumps

The performance of the simple single-stage mechanical VCHP is investigated with varying the pressure levels in the evaporator and condenser and the amount of superheating in the compressor suction line, which are the three parameters chosen to define the VCHP operation, as described in Section 6.3.1. The influence of the condensation and evaporation pressures are reported using reduced pressures:

$$P_{\mathrm{r}} = \frac{P}{P_{\mathrm{crit}}}, \tag{6.23}$$

where P_{crit} is the fluid critical pressure.

Table 6.5 Working conditions for the performance investigation of a compressor-driven heat-pump system

Description	Value
Working fluid	R152a
Heat-source temperature	$T_{ci} = 40\,°C$
Heat-source mass flowrate	$\dot{m}_c = 1$ kg/s
Heat-source specific-heat capacity	$c_{p,c} = 4{,}200$ J/kg/K
Heat-sink temperature	$T_{hi} = 80\,°C$
Heat-sink mass flowrate	$\dot{m}_h = 2$ kg/s
Heat-sink specific-heat capacity	$c_{p,h} = 4{,}200$ J/kg/K
Pinch-point temperature difference in condenser	$PP_c = 6\,°C$
Pinch-point temperature difference in evaporator	$PP_h = 6\,°C$
Compressor isentropic efficiency	$\eta_c = 80\%$

To determine how these thermodynamic parameters affect heat-pump performance, a case study is defined. As outlined in Section 6.4.1, high-GWP refrigerants are being phased down following the Kigali Amendment to the Montreal Protocol, while HFOs and HCs pose additional challenges due to their flammability. A relatively low-GWP HFC fluid is thus chosen for the current investigation. The performance of a VCHP heat-upgrading cycle using R152a as a working fluid, extracting heat from a low-temperature ($T_{ci} = 40\,°C$) waste-heat source and releasing process heat at a high-temperature ($T_{hi} = 80\,°C$) is analysed. The remaining operating conditions are summarised in Table 6.5.

First, the amount of superheating at the evaporator outlet is fixed ($\Delta T_{sh} = 6\,°C$), while its influence on the overall system operation is described later. The effect of the reduced evaporation pressure, $P_{r,evap}$, and reduced condensation pressure, $P_{r,cond}$, on the COP, ϕ, the heating power, $\dot{Q}_h$, the working-fluid mass flowrate, $\dot{m}_{wf}$, and the power consumption, $\dot{W}_c$, is determined through a parametric analysis, whereby both pressure levels are varied while checking that the pinch-points constraints in both heat exchangers are respected. The resulting performance maps are presented in Figure 6.9.

It appears clearly from Figure 6.9 that ϕ and $\dot{Q}_h$ are inversely proportional to each other, the COP being maximised at high evaporating pressures and low condensing pressures (i.e. in the bottom-right corner of Figure 6.9a), while the amount of heat delivered being maximised at high $P_{r,cond}$ and low $P_{r,evap}$ (i.e. in the top left corner of Figure 6.9b). In other words, low-pressure ratios across the compressor lead to enhanced heat-upgrading efficiency while reducing the thermal power output together with the power consumption, as shown in Figure 6.9d.

We now investigate the influence of the third thermodynamic parameter, which is the compressor suction superheat, by varying the value of ΔT_{sh}, while the condensing and evaporating pressures are fixed to 28.5 and 6.1 bar, respectively. The results of this single parametric analysis are presented in Figure 6.10. The influence of the suction superheat on the pressure–enthalpy diagram, shown in Figure 6.10a, is

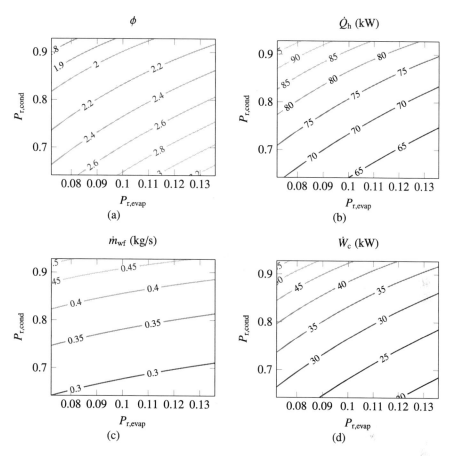

Figure 6.9 Effect of the reduced evaporation pressure, $P_{\text{r,evap}}$, and reduced condensation pressure, $P_{\text{r,cond}}$, on: (a) the coefficient of performance, ϕ; (b) the heating power, $\dot{Q}_{\text{h}}$; (c) the working-fluid mass flowrate, $\dot{m}_{\text{wf}}$; and (d) the power consumption, $\dot{W}_{\text{c}}$, of a VCHP operating under the conditions defined in Table 6.5. The amount of superheating at the evaporator exhaust is fixed: $\Delta T_{\text{sh}} = 6\,^{\circ}\text{C}$

straightforward to understand. By letting more heat into the closed heat-pump loop, that is, by increasing the enthalpy difference across the evaporator, more heat can be rejected in the high-temperature heat sink, which results in an increase of the enthalpy difference across the condenser, while the compression work remains essentially the same. The ratio between the condensation and evaporation enthalpy differences thus increases. In other words, the heat-pump coefficient of performance, ϕ, increases. For $\Delta T_{\text{sh}} = 4\,^{\circ}\text{C}$, the predicted COP equals 3.36, which increases up to 3.49 for $\Delta T_{\text{sh}} = 30\,^{\circ}\text{C}$ (i.e. increases by 3.7%). While a significant increase in the suction superheat does not lead to a major increase in performance, it has a strong influence on the discharge temperature, which increases from 115 $^{\circ}\text{C}$ up to 140 $^{\circ}\text{C}$, as seen in Figure 6.10b.

Beyond the possible oil and working-fluid degradation at high temperatures, increasing the discharge temperature has detrimental effects on the performance:(i)

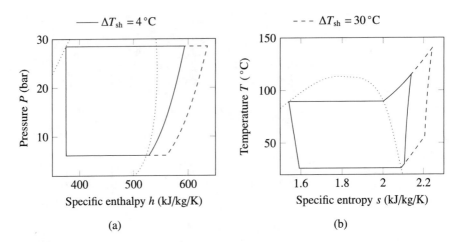

Figure 6.10 Influence of the compressor suction superheat on the T–s and P–h diagrams of a VCHP heat-upgrading cycle using R152a as a working fluid, extracting heat from a low-temperature ($T_{ci} = 40\,°C$) waste-heat source and providing high-temperature ($T_{hi} = 80\,°C$) process heat under the conditions stated in Table 6.5

as heat losses between the compressor exhaust and the condenser inlet increase; and (ii) as the exergy destruction in the condenser increases substantially. With liquid-phase heat-sink fluids, such as pressurised water, this may even cause evaporation in the heat-sink line. For those reasons, even though a small increase in first-law performance indicators is measured, it is often preferred to keep the superheat to a low value, yet not too small to ensure that the compression suction line remains liquid-free. This is the role of the metering device, that is, the expansion valve, to control the compressor suction superheating degree (typically set around 6 °C) while operating in time-varying conditions.

6.5.2   Absorption Chillers

As in the analysis of VCHPs, a case study is defined to investigate the absorption chiller performance. Due to its higher flexibility in comparison with H_2O/LiBr systems, the NH_3O/H_2O is chosen as the refrigerant/absorption medium pair for this parametric study. The assumptions and thermodynamic model described in Section 6.3.2 are used to predict the performance of an absorption chiller powered by a high-temperature heat source at T_{hi}, which ranges between 45 and 180 °C, while cooling temperatures, T_{ci}, ranging from −20 to 20 °C are considered.

A contour map of the cooling COPs, ψ, predicted by the model is shown in Figure 6.11a, along with the corresponding COP obtained from the endo-reversible analysis proposed by Velasco et al. (1997). The latter technology-agnostic first-stage assessment of the chiller performance is found to be reliable for cold-source temperatures ranging between −5 and 5 °C and hot-source temperatures between 100 and 140 °C, that is in the centre of the variable space represented in Figure 6.11. However, the cooling COP is largely overestimated outside this area.

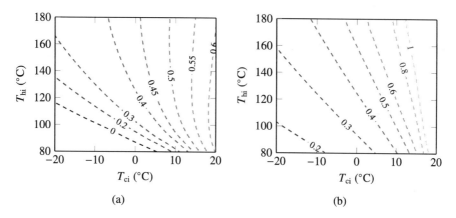

Figure 6.11 Influence of the hot heat-source temperature, T_{hi}, and of the cold heat-source temperature, T_{ci}, on the cooling coefficient of performance, ψ, calculated: (a) with the thermodynamic model of an ammonia/water absorption heat pump, as described in Section 6.3.2; and (b) with the endo-reversible formula provided by Velasco et al. (1997). The ambient heat-sink temperature is fixed at 20 °C

6.6   System Components

6.6.1   Compressors

In the previous analyses, the compressor performance is predicted solely with the isentropic efficiency, which, due to the adiabatic assumption, is reduced to the isentropic enthalpy ratio. In reality, the performance of the compression machine cannot be captured by a single parameter, even less a constant one. The compressor performance is adversely affected by various loss mechanisms that take place both within and outside the compression chamber.

Various kinds of compressors can be used in mechanical heat pumps, including positive displacement (e.g. scroll, screw and reciprocating-piston compressors) and turbo compressors. Reciprocating-piston machines are however predominantly used in compressor-driven heat pumps or chillers, the crank-driven piston technology being a mature, robust and flexible technology. The performance and use of piston compressors in heat-pumping systems is therefore discussed in more details in this section.

Reciprocating-Piston Compressors

The operating principle of a reciprocating-piston compressor is presented in Figure 6.12. A rotating motor drives the rotation of a crank, which, through a connecting rod, in turn drives the piston downwards from the top dead centre (TDC) to the bottom dead centre (BDC) and upwards back to TDC. Intake and exhaust valves are pressure-actuated reed valves. Following the course of the fluid, the first loss mechanism occurs when the intake valve opens. The upstream pressure forces the superheated vapour

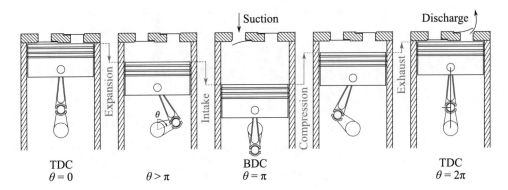

Figure 6.12 Schematic representation of a reciprocating-piston compressor working principle

in the compressor chamber through a restricted area, which entails significant pressure losses during this isenthalpic process. During the intake and compressor strokes, gas-to-wall in-cylinder heat transfer and mass leakage through the piston rings also adversely affect the compressor performance. Together with the pressure losses during the exhaust, these loss mechanisms are irreversibility sources and the reason why the compression process is non-isentropic. Along the transmission line, the friction between solid moving parts and the energy conversion losses in the motor add up the overall performance loss.

The overall compressor isentropic efficiency is defined as the ratio of the power consumed during an ideally isentropic process to the actual electrical power, $\dot{W}_{el}$:

$$\eta_{is} = \frac{\dot{m}_{wf} \, (h_{2s} - h_1)}{\dot{W}_{el}}, \tag{6.24}$$

where h_{2s} is the isentropic discharge enthalpy, as defined in Equation 6.13, and h_1 is the suction enthalpy.

Due the various and complex loss mechanisms, the compressor performance varies significantly with the operating conditions, and strongly depends on the working-fluid properties. The isentropic efficiency value can be predicted by fully comprehensive or heuristic reduced-order models based on mass, momentum and energy conservation equations, such as that proposed by Stouffs, Tazerout and Wauters (2001). These dynamic models can predict accurately the part-load performance of reciprocating engines but are computationally expensive and require tedious experimental validation.

Most compressor manufacturers however provide fluid-dependent performance maps or look-up tables, which give the expected performance as a function of both the suction and discharge temperature, assuming a given superheat and subcooling.

Fluid-Dependent Polynomial Compressor Performance Maps

The fluid-dependent compressor performance maps are given as polynomial functions and can be obtained from most manufacturers for various suction superheat levels, ΔT_{sh}. For a given rotating speed, both the consumed electrical power

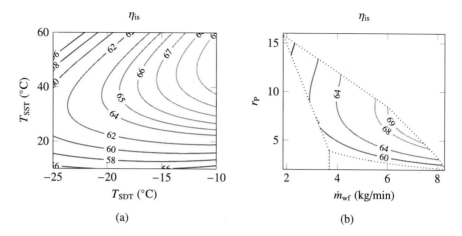

Figure 6.13 Fluid-dependent performance maps presenting the isentropic efficiency, η_{is} (%), of the Bitzer 4NES-12Y compressor operated with R134a at a fixed rotational frequency of 50 Hz, as a function of: (a) the saturation suction and saturation discharge temperatures; and (b) the pressure ratio, r_p, and working-fluid mass flowrate

and mass flowrate are given as functions of the saturation suction temperature, $T_{SST} = T_{sat}\left(P_{evap}\right)$, and saturation condensation temperature, $T_{SDT} = T_{sat}\left(P_{cond}\right)$. For example:

$$
\begin{aligned}
\dot{W}_{el} =\ & c_{w,1} + c_{w,2}\,T_{SST} + c_{w,3}\,T_{SDT} + c_{w,4}\,T_{SST}^2 + c_{w,5}\,T_{SST}\,T_{SDT} + c_{w,6}\,T_{SDT}^2 \\
& + c_{w,7}\,T_{SST}^3 + c_{w,8}\,T_{SDT}\,T_{SST}^2 + c_{w,9}\,T_{SST}\,T_{SDT}^2 + c_{w,10}\,T_{SDT}^3.
\end{aligned} \tag{6.25}
$$

In order to derive the compressor isentropic efficiency from those polynomials, a small bit of calculus is required. The compressor suction enthalpy is obtained from the saturation suction temperature and the specified superheat:

$$
h_1 = \mathrm{EoS}\left(P_{evap},\ T_{SST} + \Delta T_{sh},\ \text{fluid}\right), \tag{6.26}
$$

while h_{2s} is, by definition:

$$
h_{2s} = \mathrm{EoS}\left(P_{cond},\ s_1,\ \text{fluid}\right). \tag{6.27}
$$

As $\dot{W}_{el}$ and $\dot{m}_{wf}$ are given in polynomial forms, the isentropic efficiency is finally obtained with Equation 6.24.

A fluid-dependent performance map is thus obtained for a Bitzer reciprocating-piston semi-hermetic compressor, namely the 4NES-12Y, operated with R134a as a working fluid. The calculated istentropic efficiency is reported in Figure 6.13, both as a function of the saturation suction and saturation discharge temperatures (Figure 6.13a), and as a function of the pressure ratio, r_p, and working-fluid mass flowrate (Figure 6.13b).

These performance maps can be readily used in the compressor-driven heat pump models to account for the performance variations in part-load or off-design operating conditions.

6.6.2   Heat Exchangers

Various kinds of heat exchangers are suitable for use in mechanical or heat-driven heat pumps. As the interface between the waste-heat stream and the heat pump or chiller system, the condenser, evaporator and desorber are key components of waste-heat recovery technologies. Plate and shell-and-tube are the most common types of heat exchangers used in heat-upgrading and chiller technologies. The design, cost and performance assessment of these technologies is discussed in detail in Chapter 5, Section 5.4.4.

6.6.3   Expansion Valves: Key Metering Devices

The expansion valve placed downstream of the condenser unit is a key component of mechanical vapour-compression systems used for refrigeration, air-conditioning or for heat upgrading. The function of this flow-restricting device, often referred to as a metering device, is twofold: (i) to control the evaporator-outlet superheat to protect the compressor from liquid flooding; (ii) to control both the refrigerant mass flowrate within the closed loop and pressure ratio between the condenser and evaporator units to achieve an optimal operation of the system.

Various kinds of metering devices are used in vapour-compression systems, offering different controllability levels.

Capillary tubes are fixed-size flow-restricting orifice devices, which are commonly used in small-scale systems operating continuously under stable conditions, that is with fixed heat-source and heat-sink temperatures and constant heat loads. Despite their simplicity and low cost, capillary tubes negatively impact the heat-pump performance operating in part-load conditions and even restrict their use to a very limited range of operating conditions (outside of which the system might even have to shut down due to large amounts of liquid in the compressor suction line).

Automatic expansion valves are passive devices that maintain a constant evaporation pressure, thanks to a spring diaphragm assembly. The hotter the heat-source, the higher the evaporation pressure and the lower the mass flowrate allowed through the evaporator. For that reason, automatic spring-loaded expansion valves are ideal for applications with moderate variations in the operating temperature.

Thermostatic expansion valves (TXVs) are passive devices that maintain a constant superheat at the evaporator outlet by controlling the amount of refrigerant released into the evaporator unit. The pressure in the evaporator and the spring load apply closing forces on a diaphragm connected to the valve-opening head, while an opening force is applied on the same diaphragm by the pressure in a bulb attached to the evaporator outlet and connected to the TXV via a transmission line filled with a fixed mass of refrigerant. The latter pressure adjusts according to the actual evaporator-outlet temperature: the higher the suction line superheat, the higher the bulb pressure, and hence the larger the valve opening area. In other words, for a given evaporating pressure, if the superheat increases, the mass flowrate allowed in the evaporator increases, which in turn reduces the superheat. This self-modulating valve is commonly used in applications with significant variations in heat load and heat-source and heat-sink temperatures.

Electronic expansion valves (EXVs) offer the highest control level, with an effective opening area that can be adjusted in real time with high accuracy using a step motor connected to an electronic controller. The latter receives an input signal from a set of sensors, typically thermistors connected to the heat-source outlet, evaporator inlet and outlet to sense the superheat in the compressor suction line, while a pressure transducer can also be used for advanced custom control applications. The responsiveness of such devices allows not only for compressor floodback protection but also for advanced dynamic control, notably to optimise the instantaneous COP in highly transient applications, for example during start-up and shutdown of chiller units in transport refrigeration units, subject to frequent sudden changes in operating conditions (door opening, multiple start/stop per day). Significantly more expensive than its passive counterparts, the EXVs are preferred in custom, high-specification applications, often in conjunction with variable frequency drives to adjust the compressor rotational speed.

6.7   Summary

The different kinds of heat pumps and chillers offer great potential for a careful utilisation of waste or renewable heat sources. Both electrically driven (based on the mechanical vapour compression cycle) and heat-driven (closed-cycle thermochemical heat pumps relying on ab- and ad-sorption mechanisms) heat pumps are used for heat-upgrading or heat-removal purposes, that is, respectively for space, district, industrial process heating or for refrigeration, air-conditioning and process cooling applications.

After a quick summary of the technology-agnostic analysis presented in Chapter 4, defining a series of ideal and realistic performance indicators, the current chapter has presented reduced-order modelling tools for a refined prediction of mechanical and heat-actuated heat-pumping systems performance. The thermodynamic analysis of both systems indeed provides us with comprehensive sets of governing equations that have been used here to determine the key cycle variables (including both operating conditions and component-specific performance indicators) and their influence on the overall heat-pump performance. The main variants of the basic configurations are also presented to guide application-specific choices such as high-temperature heat-upgrading applications (typically industrial process heat), or to further improve the energy-saving potential. Moreover, the working-fluid selection (or working-fluid pairs in case of a sorption heat pump) is discussed in detail, emphasising the tension between two design objectives: (i) the development of cost-effective, high-efficiency systems; and (ii) the transition towards zero-ODP (ozone depletion potential) and low-GWP (global warming potential) working fluids.

Finally, the performance of compression-driven and heat-driven heat pumps and chillers is explored with varying operating and boundary conditions. In particular, the deviation between the technology-agnostic performance indicators and those predicted from the refined, comprehensive models is quantified to highlight the limitations of the former, which are of great value for high-level choices but need to be used carefully, especially in applications with wide ranges of operating conditions.

7 Application of Heat Conversion and Upgrading Technologies

In this chapter, the text is arranged as energy usage, recovery potential, practiced recovery technologies, barriers and opportunities for the future. The topics covered under the section in the context of industrial waste-heat recovery and heat-recovery technologies are: metals industry, chemical industry, paper mills, cement industry and food industry. Furthermore, engine waste-heat recovery, transport sector and renewable heat are also discussed.

Universally, the knowledge base of the market for 'heat' is underdeveloped. Therefore, there exist significant gaps in the information available on end uses of energy at both macroeconomic and microeconomic levels. The market for basic fuels and electricity is a widespread market, while the fuels that are associated with heat generation is not. The reason being that heat is traded as a commodity in a narrow market specifically in the process industry. Furthermore, fuels and electricity are privatised and traded in the free market, whereas heat is not. Therefore, products such as electricity and gas come as a streamlined service but not lighting and heating, as outlined by Berntsson and Åsblad (2015).

The key factors to understand and reduce harmful emissions are the market for heat and the policy options governing the heat usage. Since industry tends to use more carbon-based fuels for heating and electricity generation, harmful emissions become a significant issue, as outlined by Chintala, Kumar and Pandey (2018), Costall et al. (2015) and Berntsson and Åsblad (2015). Economically feasible waste heat recovery (WHR) applications for recovering waste heat from combustion exhaust streams with temperatures above 139 °C that are conducive is limited to specific industries and processes. For WHR flow streams with temperatures above 139 °C become economically attractive. This chapter focuses on the major industrial processes that satisfy these temperature limits. The chapter draws examples and information mainly from the two economies, the United Kingdom and the United States, to demonstrate and describe certain scenarios; however, it is generally applicable across the respective industries.

7.1 Iron and Steel

In the iron and steel industry, there is a great opportunity for heat recovery. Regardless of the technology used, many high-temperature processes are conducive for waste-heat recovery. The processes of interest are coke ovens, blast furnaces for iron production, and basic oxygen furnaces for steel production. The electric arc furnaces (EAF) that are very efficiently performed as process units are also a source of waste heat that can be captured, as shown by Alshammari et al. (2018). Other waste heat sources may

Table 7.1 Specific energy consumptions (SECs) and fuel splits.

Operation	SEC (GJ/t)	COG/BFG/NG	Solid Fuel	Electricity	Steam	Other
Coke ovens	2.95	93%		2%	5%	
Sinter strands	1.64	8%	85%	7%		
Blast furnace	14.7		75%	1%	24%	
Basic oxygen furnace	1.44	19%		39%	42%	
Continuous casting	0.31			100%		
Slab mill	2.87			36%		64%
Hot rolling	2.43			35%		65%
Cold rolling	1.69			56%		44%
Pickling	1.27			67%		33%
EAF furnace	2.5			75%		25%

Data gathered from the reports by Energetics Incorporated (2000), Entec UK Limited (2006) and the European Commission (2008)

include melting furnace exhaust, ladle preheating, core baking, hot metal pouring, shot blasting, casting cooling, heat treating and quenching. This section presents the possible sources and sinks of waste heat to be recovered and the viability of waste-heat recovery in this industry. Although this section is limited to the iron and steel sector, it should be noted that within metals manufacturing there are various other opportunities that may be technically feasible and economically viable.

The heart of the common steel production routes can be divided into two steps: primary and secondary. The primary being the blast furnace/basic oxygen furnace route. Secondary steel is via the electric arc furnace (EAF). For example, in the United Kingdom, steel is produced exclusively by these methods and the three integrated plants are Teeside, Scunthorpe and Port Talbot. A popular alternative method of steel production is direct-reduction iron (DRI). This technology is used where there is access to natural gas (NG) or refinery gas. A number of regions, including the Americas, the Middle East, Australia and Japan, have many DRI steel plants.

During primary steel production by integrated plants there are many processes that make up the whole plant operation. As an example, consider Table 7.1, in which the individual operational plant steps are listed and the way the fuel is divided between these streams is given. The specific energy consumptions (SECs) is a metric used for comparison.

The opportunities for waste-heat recovery from a typical iron and steel integrated plant are summarised in Table 7.2, and each process is described in more detail in the following subsections.

7.1.1   Coke Ovens

Coke produced from coal via pyrolysis has a dual role in the blast furnace as a reducing agent and a fuel. The coke oven's temperature is approximately 1,100 °C, and three

Table 7.2 Opportunities of WHR from process exhausts.

Process/Unit	Process Temperature (°C)	Exhaust Temperature (°C)	Recoverable Heat (GJ/trs)		Exhaust Stream Gas (g) or Solid (s)	Heat-Recovery Technology
			LOW	HIGH		
Coke ovens	1,100	1,100	0.12	0.24	Hot coke (s)	Dry quenching
Sintering	1,350	350	0.49	0.97	Cooler and exhaust gas (g)	Advanced stingering
Blast furnace	1,500	150	0.16	0.31	BF gas (g)	Top-pressure recovery turbine
Basic oxygen furnace	1,600	1,600	0.1	0.2	BOF gas (g)	Gas recovery/ boiler
Continuous casting	980	800	0.25	0.5	Cast slabs (s)	Radiant heat boilers
Hot rolling	900	900	0.31	0.62	Steel (s)	Water spraying/ heat pumps

Data collected from the review paper by de Beer, Worrell and Blok (1998)

major heat losses from coke ovens can be observed, as noted by Berntsson & Åsblad (2015):

- glowing coke at 1,100 °C, estimated radiant heat of 43–60% of the thermal energy;
- coke oven gas (COG) at 650–800 °C, estimated convective heat 20–30%;
- waste combustion gas at approximately 200 °C after partial recuperation of heat, amounting to 10–18%.

Coke is quenched to stop burn-off. Quench techniques are either a wet quench with direct water spray or a counter-current gas quench. In both cases heat is recovered by the steam generated (Costall et al. 2015; Chintala, Kumar & Pandey (2018)). Although COG streams have high enthalpy, it is an under-utilised heat source (Berntsson & Åsblad 2015). To prevent condensed tar scaling the exhaust gas temperatures are maintained above 400–450 °C. COG enthalpy is a potential heat source for waste-heat recovery. The current strategy is to preheat the fuel gas mixture.

7.1.2 Blast Furnaces

The blast furnace (BF) is the heart of an integrated steel mill, and the blast furnace in any configuration is the largest energy consumer (Costall et al. 2015). Within the blast furnace the conversion of iron ore (iron oxide, FeO) into pig iron (Fe) takes place. The blast furnace operates by a top feed transported downwards while a blast of hot air is charged upwards to meet the solid. The solid feed contains the chunks of iron ore, sinter or pellets, flux and coke. Multiple auxiliary stoves, known as Cowper, maintain hot air at a high blast temperature. The two opportunities for heat recovery are the cowper exhaust gases and the BF gases (BFG) leaving the BF.

Similar to the COG, BFG is a syngas with value components of approximately 20–28% CO, 1–5% H2, blended with inert components about 50% N2, 17–25% CO2, as well as particulates, sulphur and cyanide compounds. BF exhaust temperatures are about 400 °C. 'BFG is considered to be a fuel and is utilised as sensible heat contained in the gas (Energetics Incorporated 2000).

There is a good WHR opportunity from the exhaust gases leaving hot blast stoves. The gas temperatures are approximately 250 °C (Alshammari et al. 2018), and also clean to make WHR economically attractive.

7.1.3   Basic Oxygen Furnace

The basic oxygen furnace (BOF) is an oxidising reactor to clean up the impurities of the pig iron. Pig iron contains the impurities carbon, silicon, phosphorous, sulphur and manganese that have to be removed and controlled in the process of producing steel. The heated feed charged into the BOF is pig iron, scrap metal and oxygen with fluxes. In the furnace, the exothermic oxidation reaction maintains the high temperatures to transform the metal from solid state to liquid state.

The BOF gases are at around 1,600 °C and with a high concentration of carbon monoxide that is similar to COG and BFG, it has similar opportunities in terms of chemical energy and sensible heat. The main disadvantage in waste-heat recovery is the impurity content. The high particulates, iron oxides, heavy metals, SOx, NOx and halides pose challenges (Energetics Incorporated 2000; Alshammari et al. 2018).

In the United States, as a traditional practice driven by economics, BOF gases are flared. However, with higher heat cost and clever techniques for WHR, Europe and Japan make the overall process economically viable (Alshammari et al. 2018).

7.1.4   Electric Arc Furnaces

Electric arc furnaces are a popular method of secondary steel production. Approximately half the US steel production is based on recycled steel scrap. Electric smelting technology has made considerable advances in steel production using scrap metal.

The furnace structure is a refractory-lined vessel with a collapsible top. The gap between the tip of the carbon electrodes and the charge can be adjusted to get the proper high-voltage arc by adjusting the electrode height movement of the mount elevation. The electric arc generates the heat for the metal to melt. The EAF charge of scrap metal, direct reduced iron, hot briquetted iron and cold pig iron is transported from the top hoppers. Other additives include fluxes and alloying agents to get the desired quality of the product (Alshammari et al. 2018).

At peak loads the EAF off-gas temperatures reach in the range of 1370–1925 °C. The energy losses are estimated around 20% of the total energy input. Furthermore, the chemical energy of the off-gases is comparable with the amount of sensible heat lost. This opportunity is well recognised, and, as a result, various technologies are used for WHR, of which currently the most common practice is feed preheat (Energetics Incorporated 2000; Alshammari et al. 2018).

Table 7.3 Opportunities of WHR from hot solid streams in iron and steel production.

Waste-Heat Source	Maximum Temperature (°C)	Sensible Heat (MJ/t)	Applicable Steel Production (Mt/year)	Recovery Technology	Waste Heat (PJ/year)
Hot coke	1,100	0.22	56.5	Dry coke quenching	12.7
Blast furnace slag	1,300	0.36	56.5	Radiant heat boiler	20
Basic oxygen furnace	1,500	0.02	56.5	Radiant heat boiler	1.1
Cast steel	1,600	1.3	104.6	Radiant heat boiler	132
Hot rolled steel	900	5	104.6	Water spraying/ heat pumps	524

Data collected from the report by BCS Incorporated (2008)

7.1.5   Waste Heat from Solid Streams

There are significant waste-heat recovery opportunities in solid streams and cooling water streams. The significant solid streams are hot cokes, BF by-product fuels slag, BOF slag, cast steel and hot-rolled steel. A representative summary of heat losses of selected systems is provided in Table 7.3.

In the iron and steel industry, there are substantial opportunities of recovering significant amounts of heat in the solid streams. WHR from the solid streams represents a significant engineering challenge and a recognisable capital investment. It is logical to predict that with the cost of energy and the concerns of the CO_2 footprint incentivises the drivers to tip the balance in favour of serious WHR.

The opportunities for harnessing the sensible heat from coke, BF and BOF slag become apparent because of the significant amounts of sensible heat. In countries such as Japan and Germany, the technology is proved and mature, although wide application of technology to WHR has been limited. It is changing with the social attitudes and the responses to the price increase of energy and the environmental concerns.

WHR opportunities also exist in the cast steel and hot-rolling processes. The plot plan and the operational footprint influences the WHR; therefore, brown-field retrofits can be an engineering challenge. However, for Greenfield plants WHR becomes attractive because of the degrees of freedom with the inside battery limits.

7.2   Aluminium Manufacture

In the aluminium industry, waste-heat recovery can again be broadly divided into two sections, namely: primary production and secondary melting processes. There are other manufacturing processes related to aluminium industry, such as packaging and engineering metals; however, smelting is the predominant energy-intensive process.

The secondary aluminium production is insignificant because of the relatively smaller energy requirements, which is typically 5% of that required for primary production (Berntsson & Åsblad 2015).

Modern aluminium smelting processes use only prebaked anodes. Aluminium manufacturing process includes the electrolysis of alumina (aluminium oxide), derived from bauxite, and is dissolved in a bath of sodium aluminium fluoride (cryolite) at an approximate temperature of 1,000 °C (Berntsson & Åsblad 2015; Costall et al. 2015; Alshammari et al. 2018).

The highest energy consuming process is electrolysis, but there are other significant on-site energy demands such as casting. Exhaust gases from the process are tapped from the reduction cells filtered to comply with environmental legislation and then emitted to the atmosphere at a temperature around 100 °C. Half of the overall input energy is lost as heat, and 30% is in the enthalpy of the off-gas. The key technical constraint of the sensible heat recovery in these gases is fouling.

7.2.1 Primary Production

The WHR stages are briefly described here. Refining the ore, smelting, anode baking and melting are the most significant concerns of heat recovery because of the thermal energy intensity and the high-temperature operations.

7.2.2 Refining

A significant fraction of the total energy required to produce aluminium is spent in the calcination process. Waste-heat exhaust from calciners is a focal point for waste-heat recovery. However, the exhaust contains high moisture, in the range of 50% by volume, and contaminants as particulates, especially alumina, at a low temperature (185–200 °C).

Waste-heat recovery from exhaust gases and hot alumina at low temperatures and with the presence of particulates pose an engineering challenge. Regardless of the type of heat exchanger, severe fouling becomes an issue. A multitude of technologies can be considered to recover latent heat from the water vapour released by the industry (Bisio & Rubatto 2000; Alshammari et al. 2018).

Another excellent source of WHR is from hot alumina discharged at a very high temperature. However, the downside is that the particulates present an operational difficulty. Coke calciner afterburners also represent possible waste-heat source, with gas streams with temperatures of about 1,090 °C that could be used to produce steam or electricity.

7.2.3 Smelting

The electrolysis pots are also a heat recovery opportunity. The radiative heat losses from the 300–550 °C vessel walls are about 20% and the convective heat losses from the off-gases are estimated to around 15% (Berntsson & Åsblad 2015; Alshammari et al. 2018). Fugitive emissions are reduced by make-up air, and the duct system needs to be opened for periodic maintenance. Cleaning gases remains a challenge (Alshammari et al. 2018).

7.2.4   Anode Baking

The carbon anode is an important component in the aluminium smelting process. It is also the most energy-intense component, and therefore the most expensive operational stage. The anode backing energy is derived from fuel combustion. WHR opportunities exist surrounding the combustion processes, particularly around both the radiative and convective heat losses of the process. Exhaust gases are around 400 °C, making WHR favourable. However, the challenges are associated with the large excess air in the system (Berntsson & Åsblad 2015; Costall et al. 2015; Alshammari et al. 2018).

7.2.5   Melting

There are two cast sides: primary and secondary. The primary cast furnaces are semi-continuous and mostly heated by gas firing. The exhaust and temperatures therefore are not stable and fluctuate. The potential WHR sources are the wall surfaces, openings and the exhaust gases. However, these opportunities are considered as low impact and complex challenges (Berntsson & Åsblad 2015; Costall et al. 2015; Alshammari et al. 2018).

The high-temperature exhaust gases from the secondary-side melting furnaces are a WHR opportunity. The exhaust with a temperature greater than 870 °C releases a significant heat. However, the possibility of impurities such as particulates, organic vapours, and flux material vapours picked up from the melting furnace becomes an engineering challenge.

7.2.6   Recycling and Secondary Melting

Exhaust gases from reverb, crucible heater and gas-powered generators also represent WHR opportunities. The relatively clean exhaust, with temperatures ranging between 790 and 1,090 °C, becomes attractive. Preheating combustion air and low-temperature power generation via organic Rankine cycle (ORC) methods are possible. The engineering challenge is to address the fluctuating mass flow rates (Berntsson & Åsblad 2015; Costall et al. 2015; Alshammari et al. 2018).

Another WHR opportunity is in the area of surface cleaning of used cans. This involves de-lacquering, heating and incineration processes that have a fairly clean and consistent exhaust mass flow at a temperature of around 480 °C. With modern WHR techniques, the economics could be viable to be accepted into a business model (Costall et al. 2015; Alshammari et al. 2018).

Crucible heaters can be an opportunity for WHR with modern and cleaver engineering. Although the units are small, the exit temperatures are around 790 °C, making the heat recovery attractive (Berntsson & Åsblad 2015; Costall et al. 2015; Alshammari et al. 2018).

7.3   Chemical Industry

There are various opportunities for WHR in the chemical industry. In this sector, direct process heat losses from surfaces as radiative heat and convective heat losses via gaseous streams are very significant. The most promising opportunities in the chemical sector are described next.

7.3.1   Ammonia

Ammonia is the base chemical for almost all synthetic nitrogen fertilisers produced in the world. It is produced by reacting nitrogen and hydrogen by the Haber or Haber–Bosch process. Hydrogen is produced through steam reforming of hydrocarbon feedstock. Almost the same proportion of ammonia is used in the production of nitrogen-based synthetic fertilisers (Berntsson & Åsblad 2015; Alshammari et al. 2018). Ammonia synthesis is an exothermic reaction. The net energy input in the overall ammonia production process is in the manufacture of hydrogen, which typically occurs in a two-stage reforming process at temperatures up to 1,000 °C. In modern ammonia plants most of the exhaust heat from the reformers and ammonia synthesis process is recovered by massively heat-integrated designs with clever engineering. However, with the increasing energy cost there is an opportunity for more WHR by a high degree of overall plant integration. More waste heat from the reformers is recoverable with modern design that can justify the capital invested (Berntsson & Åsblad 2015).

7.3.2   Ethylene Furnaces

Ethylene is the largest volume petrochemical product in the world and is a main building block for many consumer products. Over 150 million tons of ethylene was produced in 2018, requiring about 3,587 PJ of energy input. A fundamental process unit of the ethylene production process is the pyrolysis furnace, where hydrocarbon feed is cracked at high temperatures around 760–870 °C (Alshammari et al. 2018).

Although the modern crackers are excellent integrated units, there still exist opportunities for waste-heat recovery within ethylene furnaces. Especially, the cracker flue gas exiting the convection section is an attractive opportunity. Although the exit gases are at relatively low temperatures, typically around 150 °C, with the rising cost of energy and environmental compliance, this could represent an economically feasible waste-heat recovery opportunity (Alshammari et al. 2018).

7.3.3 Petroleum Refining

Essential refinery unit operations include distillation, cracking, reforming and hydro-treatment. All of these processes are energy-intensive, high-temperature and exothermic processes. Although there is heat integration, there are many operations that release high-quality waste heat that is not recovered, which could be an opportunity for WHR (Costall et al. 2015; Alshammari et al. 2018).

WHR opportunities from the primary chemical and petrochemical operations include exhaust gases from all fired heating systems, including crackers, and from power generation, gas turbine and heat recovery steam generator equipment.

The use of NG or liquid petroleum gas as fuel means that the exhaust gases are clean, whilst the typical waste heat discharged ranges from 95 to 180 °C (Alshammari et al. 2018). Exhaust gases from thermal oxidisers (TOs) may also be a good opportunity. The temperatures of these exhaust gases are favourable and are as high as 760–980 °C. The down side is that they are can have chlorinated compounds and other inorganic contaminants, posing engineering challengers with corrosion.

Another WHR opportunity is to collect stack gases and use them to heat other streams. Although, collecting streams from multiple stacks is a significant logistical effort and retrofitting heat exchange is not economically attractive.

In refineries and petrochemical plants, heat is wasted in large quantities as low-pressure steam discharge. With the current technology, energy recovery from low pressure is not economically feasible. The development of reliable disposal of excess gases that are flared could result in energy savings. However, to include WHR of flared gases in the business models, the capital cost and the operational cost have to be justified with the energy cost.

7.3.4 Other Major Chemicals

There are three major chemicals that are most energy intense to manufactures: ammonia, chlorine and ethylene. These take up approximately 23–34% of total chemicals sector energy consumption (Berntsson & Åsblad 2015; Costall et al. 2015; Alshammari et al. 2018). Other chemical processes that are less-energy intense but significant are titanium dioxide and soda ash production. For example, in the United Kingdom in 2006, the production of ammonia, chlorine, ethylene, soda ash and titanium dioxide accounted for around 31–42% of the chemical sector's total energy consumption (Berntsson & Åsblad 2015). In the soda ash industry, the process temperatures average around 3,000 °C and exhaust reach 1,500 °C, whilst in the titanium dioxide processing the process, temperatures reach 1,200 °C, and exhaust temperatures reach an average of 2,500 °C (Berntsson & Åsblad 2015). The estimated potential for heat recovery from these processes could be 5–10% (Berntsson & Åsblad 2015). With the current heat prices and the environmental footprint pressures, there might be good potential for heat recovery. Furthermore, it could be anticipated that with novel technologies an attractive business case could be demonstrated.

7.3.5   Paper Mills

The expectation of energy efficiency improvements in this sector is dependent on advances in technologies. With the state of the art dryers, it is not expected beyond 10% improvement by waste-heat recovery at hot water discharges and emissions from drying processes (Berntsson & Åsblad 2015; Costall et al. 2015).

Issues that are encountered in waste heat recovery in this industry are low temperatures of exhaust streams, high humidity and the airborne particulate matter. Although there exists an engineering challenge, this presents an opportunity for developing a novel system for heat recovery (Alshammari et al. 2018).

7.3.6   Vinyl Coating Mills

Ovens heat vinyl coatings with different substances to cure the material. The ducts from multiple controlled temperature zones are combined into a main duct that is connected to one or more regenerative thermal oxidizers (RTOs). The material is heated to about 200 °C and then cooled before the exit. The waste heat is in two forms: chemical heat of the generated volatile organic compound (VOCs) and sensible heat of air.

Similarly, dryers, RTOs, steam boilers and oil heaters are associated with exhaust gas streams that operate at higher temperatures even up to 300 °C that can be viable sources for WHR. VOCs are an inconvenience that can make the waste-heat recovery economically feasible.

7.3.7   Ceramic Industry

The ceramic industry is one of the most energy-intensive industries where both electrical energy and chemical energy are used. Electrical energy is used to power the motors and the chemical energy to heat the kilns and furnaces. In general, the ceramic production process typically consists of five stages.

The raw material with additives is formed into a slurry and then converted to a damp powder to be pressed to shape. Drying by controlled heating further reduces water before it enters a kiln to be fired to make the ceramic. This is followed by surface finishing. Drying and firing operations are the two top energy-consuming processes and are associated with the highest emissions. The largest energy consumer is at the firing stage, during which temperatures can reach up to 1,800 °C. Many clever techniques are used for WHR during the drying and firing steps. The most successful technology is the WHR by cogeneration with associated organic Rankine cycles (ORCs) and heat pipe systems (Bisio & Rubatto 2000; Alshammari et al. 2018). Economisers may also be used to preheat the combustion air to perform kiln WHR (Berntsson & Åsblad 2015; Costall et al. 2015).

In the brick manufacturing industry, temperature within the firing zone reaches around 1,000 °C. WHR from the kiln is used in the dryer. However, about 35–40% of the total heat input is rejected via exhaust at an approximate low temperature, 150 °C. The overall WHR from the combustion emissions in brick manufacture is around 50%

(Berntsson & Åsblad 2015). With novel technology and smart engineering, the overall efficiencies could even be driven further.

7.3.8   Cement Industry

The primary sources of waste heat from common cement plants that use the dry process are clinker process, exhaust gases and the kilns.

The clinker cooling air is a WHR opportunity. The airflow temperature it is around 480 °C. WHR technologies use air preheating of cold streams. Exhaust gases from the system and the preheaters are excellent point sources. These gases contain a high percentage of CO_2 and are contaminated with combustion products from fuels. The gas temperatures are relatively low and can fluctuate from 150 to 260 °C (Alshammari et al. 2018).

Heated surfaces of kilns and high surface temperatures of pre-calciners are up to about 430 °C. There are no economically viable retrofit technologies to recover this heat yet but with further development, it is a possibility. The best solution is to Greenfield design for the insulation-refractory system to reduce the kiln shell temperature (Alshammari et al. 2018).

Four common manufacturing methods exist in the cement industry: wet, semi-wet, semi-dry and dry. The current preference is the dry process because it is less energy intense; however, the technology choice depends on the raw materials. The common process of cement manufacture involves calcining of calcium carbonate from limestone at a temperature of around 1,000 °C to produce calcium oxide. The clinkering process then occurs at around 1,500 °C, when the calcium oxide reacts with silica, alumina and iron oxide to form the silicates, aluminates and ferrites of calcium, which comprise clinker. The clinker is then ground and blended with gypsum and other additives to produce saleable cement.

The wet and dry processes differs with the moisture content. The heat from the back end of the kiln is used to preheat the feed to desired temperatures. Then a precalciner initiates and carries out the majority of the calcination process before the kiln. It is estimated that the preheater kiln energy demand is about 60–40% (Costall et al. 2015). By forcing cooler air to contact with the hot clinker, the recovered heat is recycled into the process (Berntsson & Åsblad 2015).

There are two exhaust streams from a typical cement plant: from the precalciner or preheater at a temperature around 200–300 °C and the other stream from the cooler at around 300–400 °C. It is estimated that these two exhaust streams combined is about 25–35% of the total heat input. The major heat losses are through the preheater/precalciner stack (Costall et al. 2015). There could be further opportunities for more WHR with novel technologies and high cost of heat.

7.3.9   Fibreglass and Glass Manufacturing

Energy-use systems in the industry are very similar in regard to waste-heat discharge and their characteristics. Glass melter devices use NG-fired burners with preheated air or oxy-fuel burners, electricity or hybrid system. Exhaust gases from the furnace and channels are excellent sources of WHR. These gases are at very high temperatures, reaching beyond 1,370 °C. However, the challenge is to manage the contamination of source streams with inorganic particles and condensed vapours of elements at a specific temperature range. The gases from the system can be around 480 °C, while gases exiting the baghouses can be as high as 230 °C (Alshammari et al. 2018).

In fibreising process, a thermal energy-intense and large quantity of NG is used to maintain the required high temperatures. This stage can be a potential WHR opportunity. The challenge is that the exhaust streams are consolidated into one stream and thus the exit temperatures tend to be as low as 260 °C (Alshammari et al. 2018). The WHR and application as space heat can be a strategic advantage. The target WHR projects can be well placed and economically justified in an attractive business model.

7.4   Food Manufacturing Industry

Within the food processing and manufacturing sector, the main thermal energy activities are associated with boilers for raising steam. Therefore, the main waste-heat sources are surrounding the boiler systems and steam utilisation.

The waste heat in this industry is at low-to-moderate temperatures of 150–350 °C, with a varying moisture content. The main heat sources are the boiler exhaust gases, oven gases and the dryers. There exists a good opportunity of WHR combustion air preheating via economisers and other heat exchangers. Although retrofits may be an engineering challenge with the cost of heat, there is promise for WHR with economic feasibility.

WHR from heated water is an opportunity that has to be further explored to justify a business case. There are technologies such as ORC to harness waste heat from relatively low-temperature heat sources that could be deployed if economics allow.

In the meat processing industry, in addition to the hot water heat losses, there is another source of WHR in refrigeration (Bisio & Rubatto 2000). Pasteurisation and refrigeration are the key heat sources for WHR in the dairy processing and are followed by the dryer exhaust. Well-planned WHR systems with economisers or CO2 heat pumps can have great opportunity (Brueckner et al. 2014). Furthermore, similar opportunities exist in the egg-processing industry and the canning industry.

In the food processing industry, with the objective of low-temperature heat sources, thermo-acoustic heat engines have also been suggested. This technology converts thermal energy in to acoustic energy (Campana et al. 2013b).

7.5   Engine Waste-Heat Recovery

When considering engine waste-heat recovery, there are two key issues that need to be addressed: (i) fuel prices, and (ii) harmful emissions. Considering fuel economics and the environmental footprint, the thermal efficiency of the engine is of paramount importance. The idea of increasing the thermal efficiency using 'outside the cylinder techniques' is now widely considered, such as converting the engine waste heat into a useful form, such as mechanical or electrical power.

7.5.1 Stationary Engine Waste-Heat Recovery

The idea of renewable energy has created new challenges in production, transmission and distribution of energy. To replace traditional central energy production, decentralised energy production is conceptualised, which opens up additional opportunities for waste-heat recovery.

In an internal combustion (IC) engine, about 25% chemical energy of the fuel is converted into mechanical energy. The remaining 75% of the chemical energy is lost as heat energy via exhaust, coolant and radiation. Within these engines, efforts have focused on recovering the waste heat by capturing the unused heat using Rankine cycle. Improvements demonstrated are about 12% in power and about 15% in fuel economy (Brueckner et al. 2014). Other forms of engine waste-heat recovery that are popular include the following:

- electric turbo-compounding systems;
- thermodynamic ORC;
- thermoelectric generators (TEG);
- hydrogen generation by using exhaust gas heat energy;
- hybrid pneumatic power systems (HPPS).

The HPPS system is able to achieve the best improvement in fuel economy out of the technologies. HPPS is the most attractive WHR technology for vehicle engines; TEG is favourable and competitive for working conditions such as power plant and marine engine.

As a viable solution to residential-scale power production at low cost, high-efficiency generators for small electrical and thermal systems (GENSETS) are considered. For the efficiency improvement of small IC generators, a WHR bottoming cycle is coupled to gain about 7% efficiency advantage.

7.5.2 Transport Sector

It is known that using Rankine cycles for WHR has the potential to improve the overall engine efficiency. Alongside waste-heat recovery from stationary engines, there is also a large amount of research and development into technologies suitable for recovering waste heat from engines used within the transport sector, with applications of interest including lightweight and heavy-duty vehicles alongside marine engines.

The most widely used technology employed for this purpose is probably the turbocharger. A turbocharger uses the energy contained within the high-temperature gas leaving the engine to generate power, which, in turn, drives a compressor that increases the pressure of the air entering the engine and thus increases the air intake. However, with the increasing need to reduce our environmental impact and improve energy efficiency additional technologies to further increase efficiency are being sought. Other technologies that are also being considered include TEGs and heat pipes (Orr et al. 2016).

One of the primary technologies under investigation is the Rankine cycle, or ORC, which can make use of both the engine cooling water and the high-temperature engine exhaust to generate additional power (Wang et al. 2011). For example, it has been suggested that a properly designed ORC for waste-heat recovery from an engine exhaust could reduce emissions and pay for itself through fuel savings in approximately two to five years (Sprouse & Depcik 2013). Within the automotive sector, the low thermal loads available mean that the ORC systems are inevitably of a low power rating (in the range of a few kilowatts up to tens of kilowatts), which introduces significant challenges with regards to the design of the expanders used within these systems. This coupled with the need to balance performance against the size, weight and cost of the ORC system, alongside the transient nature of engine operation, means that the design of these systems is complex. As a result, ORC systems for this application have yet to be successfully commercialised, but remain an active area of research and development in terms of expander development, system design and optimisation (Uusitalo et al. 2013; Costall et al. 2015; Alshammari et al. 2018) and control (Esposito et al. 2015; Tona & Peralez 2015). The challenges around size, weight and transient operation are perhaps slightly mitigated when moving from lightweight vehicles to heavy-duty vehicles; heavy-duty vehicles have more space available for the integration of an ORC system, and spend a longer proportion of their time operating under steady-state conditions. Thus, this application may represent the more promising option from the point of view of initial commercialisation.

Moving up the size scale, waste-heat recovery is also being considered for waste-heat recovery marine engines, and review of potential options is explored by Shu et al. (2013). In particular, ORC systems are also being investigated for this purpose (Larsen et al. 2013; Song, Song & Gu 2015). Compared to automotive waste-heat recovery, these engines are associated with significantly higher thermal loads, and also represent less transient operating conditions. Thus, these systems are more similar to those being developed and installed for WHR from stationary engines.

7.6   Renewable Heat Sources

Renewable heat is defined as the heat generated by renewable sources. Examples include renewable biofuels, solar heating, geothermal and recovered waste heat. As a general rule, closer to the equator less energy is used for heating. In 2005, the United Kingdom used 354 TWh of electric power and for heating used 907 TWh, of which 81% was generated by gas (McKenna 2009; Campana et al. 2013b). Approximately half of the total energy consumed in the United Kingdom was as heat and half of that heat was used for space heating.

Regardless of the use of energy for heat, electricity or for transport, there is no competition to fossil fuels without any carbon valorisation or government subsidy.

Therefore, the market for renewable electricity and renewable heat will depend on a country sustainability outlook.

There are two models to promote renewables. In Sweden, Denmark and Finland, the government intervenes and valorises carbon. Based on the valorised results, a carbon tax and an energy tax is imposed. As an outcome, renewable heat component contributed to the final energy consumption increases.

In Germany, Spain, the United States, and the United Kingdom, varying levels of government support as a function of the technology are practised. The results show a lower renewable addition to energy consumption (Campana et al. 2013b).

7.6.1   Solar Thermal Energy

In solar thermal technology, solar energy in the form of heat is directly captured and utilised. In the passive solar building design, windows, walls and floors are made to capture, store and also distribute solar energy as purely heat in the cold seasons and then reject solar heat in the warmer seasons.

The difference in a passive solar design compared to active solar heating systems is not having intermediate devices, either mechanical or electrical. A passive solar design is highly location-specific. The design basis is based on the local climate knowledge and the weather patterns. The key components and their positions of the building such as window design and geometry, coatings, thermal insulation, thermal mass and shading will be custom-engineered. Although new building with passive solar design is preferred, clever engineering retrofits are successfully demonstrated.

Active or hybrid solar hot water systems use pumps or fans to force-circulate fluid. Often a mixture of water and glycol as anti-freeze or air is used to make the system efficient. Active systems with advanced controls maximise the system effectiveness (Pehnt et al. 2010; International Energy Agency 2011). The disadvantages of active solar systems are dependence on electrical power and equipment maintenance costs.

Solar tracking technology, both active and passive, increases the effectiveness of the solar thermal heat capture systems. However, the active systems depend on external electrical power and come with higher maintenance costs (International Energy Agency 2011).

7.6.2   Geothermal

Geothermal heating is defined as the utilisation of geothermal energy for heating applications. Although the world has a geothermal installed capacity of about 28 GW, as a percentage it is barely 0.07% of global energy consumption. The key advantage of the technology is that the overall thermal efficiency is high because of the direct energy utilisation (Element Energy Limited 2014; Papapetrou et al. 2018; U.S. Office of Energy Efficiency and Renewable Energy 2020).

Geothermal energy is accumulated by radioactive decay from the inception of the planet. High-temperature geothermal heat is transported from pockets close to the

Earth's surface. A common technology is a heat pump used to extract the heat from a heat sink (Element Energy Limited 2014).

Geothermal heat is cheap. Although, most of the geothermal heat harnessed is used for space heating, there is a range of industrial processes that utilise this heat energy, especially in desalination, domestic hot water and many agricultural applications. Iceland also uses geothermal energy to de-ice roads and pavements (Papapetrou et al. 2018).

The key to make geothermal systems attractive is to focus on economies of scale. Therefore, for space heating multiple buildings are webbed together to network whole communities. This technology is used across the world from Reykjavík, Iceland to Boise, Idaho where geothermal resources conducive and referred to as district heating.

Geothermal heating compared to geothermal electric generation is more efficient although electric power dispatch ability and transmission is far superior. The uniqueness of geothermal energy is that it is renewable and conserves natural resources. In addition, a clever engineered system has the advantage of low maintenance cost and high reliability (Element Energy Limited 2014).

7.6.3   Biomass

When biological material is used to generate heat energy, the process is called a biomass heat system. This is a general reference because biomass conversion technologies include direct combustion, gasification (pyrolysis), combined heat and power and anaerobic and aerobic digestion.

From the 1970s oil embargo to 2003 oil price increase is a stark reminder to the world of fossil oil dependency from transport to space heating. These fossil oil price shocks have compelled the world to think of sustainability and conservation. Price increases of oil, NG and coal have increased the interest in biomass for heat generation (Orr et al. 2016).

The concept of carbon balance is very attractive. The fixed carbon is used to generate the heat and also free the carbon into the atmosphere as carbon dioxide. Then, this carbon dioxide is re-absorbed by the plant from the atmosphere and fixed as plant matter within a reasonable time. However, the major disadvantage is the low energy density of biomass in comparison with the high energy density of fossil fuel (Orr et al. 2016).

It is also worth noting that biomass production competes against agricultural food production.

Biomass combustion generates carbon dioxide as an artificial de-carbon fixing step that is not a part of the nature carbon cycle. The time lag between the generation of carbon dioxide by combustion, which is de-fixing carbon to releasing it to the atmosphere, and then fixing this carbon and putting it back to the soil is the carbon debt.

Biomass utilisation for heat has carbon impact on the environment; however, the magnitude of the impact will also depend on the scale. It has advantages as a supplemental source of heat production although it comes with the same burden of air pollution risks as conventional fuels (Orr et al. 2016).

7.7   Summary and Conclusions

The choice of whether a certain waste or renewable heat is to be recovered, and the heat recovery method to be utilised, will depend on some critical factors such as temperature, phase and chemical composition of the source or the sink, as well as the nature of the desired end use for the recovered heat.

Table 7.4 is a summary of conventional heat exchange technologies according to operational temperature domains, heat sources and sinks, end uses, heat exchange technologies, dehumidification, allowable temperature differences, susceptibility to contamination and corrosivity of gases. Table 7.4 also summarises the adaptability of various recovery technologies in different applications.

Efforts have been made to installed waste-heat recovery technologies in industrial sectors such as in basic metals, chemical industry, non-metallic minerals, food and pulp and paper. Furthermore, there is a renewed interest in more advanced waste-heat recovery technologies, especially with the increase cost of energy, the pressure of global warming and the advancement of technology. The interest in alternative methods to the traditional direct reuse are also considered.

Table 7.4 Summary of current WHR technologies for various sources

Temperature Classification	Waste-Heat Source	Characteristics	Commercial Waste Heat to Power
High >650°C	• Furnaces: – Steel electric arc – Steel heating – Basic oxygen – Aluminium reverberator – Copper reverberator – Nickel refinery – Copper refinery – Glass melting • Iron cupolas • Coke ovens • Fume incinerators/flares • Hydrogen plants	• High-quality heat • High heat transfer • High-power generation efficiency • Chemical and mechanical contaminants	Waste-heat boilers and steam turbines
Medium 300–650 °C	• Prime-mover exhaust: – Gas turbines – Reciprocating engines • Heat treating furnaces • Ovens: – Drying – Baking – Curing • Cement kilns	• Medium-power generation efficiency • Chemical and mechanical contaminants	• Waste-heat boilers and steam turbines • Organic Rankine cycle (<450 °C) • Kalina cycle (<550 °C)

(cont.)

Table 7.4 *(cont.)*

Temperature Classification	Waste-Heat Source	Characteristics	Commercial Waste Heat to Power

Low <300 °C	• Boilers
	• Ethylene furnaces
	• Steam condensate
	• Cooling water of:
	  – Annealing furnaces
	  – Gas compressors
	  – Furnace outlets
	  – IC engines
	  – Refrigeration condensers
	• Hot-temperature ovens
	• Hot process liquids and solids
		• Energy contained in numerous small sources
		• Low-power generation efficiency
		• Recovery of combustion streams limited due to acidic concentration if temperature reduces below 120 °C
			• Organic Rankine cycle (>150 °C for gas and >80 °C for liquid streams)
			• Kalina cycle (>90 °C)

Obstacles to waste-heat recovery are, however, very common. As an investment strategy, profit from waste heat will not be the top business case for manufacturing companies. However, regulatory compliance will take precedence. There are major engineering challenges and key constrains to efficient heat recovery. The objective is to mitigate our impact on both human health and the environment. In all sectors of industry, harmful emission occurs in heat energy production and heat energy utilisation. In heat energy production, the expectation to reduce the latter is to use alternative sources such as renewables. However, in the area of energy utilisation, losses are generally in the form of waste heat. Therefore, waste heat recovery becomes a key issue in the pursuit to reduce our irreversible impact on the air quality and environment.

8 Thermal Energy Storage

The availability and grade of renewable energy sources (e.g. solar or geothermal heat) exhibit significant variations over short- (e.g. due to day–night cycles) or long-term (seasonal) time scales. A similar intermittency is observed in waste-heat supply in industrial or domestic environments, which triggers interest in thermal energy storage (TES) systems.

Whether arising from climatic variations or due to the dynamics of heat-rejecting industrial processes, fluctuations in heat-source temperature and power entail two main detrimental effects: (i) a mismatch between the energy supply (e.g. heat delivered by a heat pump or power generated by an ORC engine) and demand, and (ii) a degradation of the heat-utilisation technology performance due to transient effects. Acting as a buffer between the time-varying heat source and the heat-recovery system, a storage device can help cope with these dynamic constraints.

Thermal energy storage systems capture energy from a heat source by heating up a storage substance across a heat exchange surface, so that the heat stored can be later recovered by a heat sink, as shown in Figure 8.1. Thermal storage devices can also be used to store electrical energy when coupled with heat pumps and/or heat engines to convert between electricity and heat. This chapter however does not describe this typical application, referred to as pumped heat or pumped TES, but focuses on the thermodynamic performance and characteristics of TES systems. In particular, the aim is to provide a list of indicators and modelling techniques to assess the techno-economic potential of various thermal-storage media and strategies.

Thermal energy storage systems can be used either to: (i) to store excess heat and dissipate it for heating, cooling (e.g. via a heat-driven chiller) or power generation (e.g. using an ORC engine), or (ii) to absorb excess cooling power (by cooling down the storage medium) and return it for controlled cooling purposes. These two modes – respectively referred to as *heating* and *cooling* modes – share the same operating principle: a storage medium is heated up during a charging phase (storing excess heat in *heating* mode or delivering a cooling capacity in *cooling* mode), and cooled down during a discharging phase (returning stored heat in *heating* mode or absorbing extra cooling power in *cooling* mode). For clarity purposes, the thermodynamic analysis performed in this chapter is based on a TES system operated in *heating* mode. Similar results can be easily obtained for cold-storage systems by reversing the heat-source and heat-sink terms.

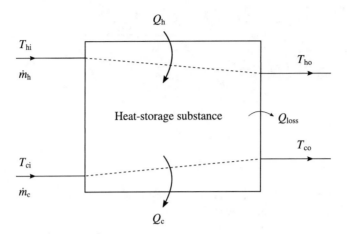

Figure 8.1 Thermal energy storage system working principle for *heating* purposes.

8.1   Selection and Design of a Thermal Energy Storage System

The principle of TES is to transfer heat to and from a heat-storage substance to either increase the efficiency of heat-utilising processes or to balance a mismatch between heat demand and supply. A distinction is to be made between heat-storage substances (or media) that involve, or not, chemical reactions. Chemical thermal energy storage (C-TES) refers to heat storage using thermo-chemical materials that capture and release heat through reversible exo- and endothermic reactions, for example, the reversible decomposition of potassium oxide into potassium peroxide and potassium. Non-chemical heat-storage media, on the other hand, store energy either as sensible heat or latent heat, depending on whether phase-change processes are involved. It must be noted that phase-change materials (PCMs) offer both sensible and latent heat storages (LHS) while they usually exhibit low-heat capacities.

8.1.1   Selection Criteria and Performance Indicators

The optimal selection, design and sizing of a TES system is highly application-dependent, due to the variety in nature and scale of the constraints. As summarised in Table 8.1, various indicators can be defined to characterise the capacity, operating range and performance of a TES system. TES optimisation processes thus require to define and prioritise selection criteria, so as to short-list adequate heat-storage media prior to designing and sizing the heat battery.

This section presents a non-exhaustive list of key indicators to evaluate the techno-economic performance of TES systems.

Size and storage capacity are fundamental characteristics of a TES system. Whether chemically reacting substances, PCMs or single-phase heat-storage substances are used, the ratio of the energy storage capacity, Q_{st}, to the corresponding storage media volume, V, is a key parameter in the choice of the heat-storage type. Defined

Table 8.1 Key characteristics and performance indicators of TES systems

Description	Indicator	Unit
Storage-medium characteristics		
Storage capacity	Q_{st}	kWh
Exergy storage capacity	X_{st}	kWh
Heat-storage substance volume	V	m^3
Energy-storage density	$q = Q_{st}/V$	kWh/m^3
Storage temperature	T_{st}	°C
Spatial footprint	A/Q_{st}	m^2/kWh
Thermal inertia (effusivity)	e_{th}	$J/m^2/s^{1/2}/K$
Thermal insulation	G_{ins}	W/K
Thermodynamic performance		
Nominal charging/discharging rates	$\dot{Q}_{h/c}$	kW
Charging efficiency	$\eta_h = Q_{st}/Q_h$	-
Discharging efficiency	$\eta_c = Q_c/Q_{st}$	-
Storage efficiency	$\eta_{st} = Q_c/Q_h$	-
Charging exergetic efficiency	$\eta_{exg,h} = X_{st}/X_h$	-
Discharging exergetic efficiency	$\eta_{exg,c} = X_c/X_{st}$	-
Exergetic storage efficiency	$\eta_{exg,st} = X_c/X_h$	-
Techno-economic assessment		
Cost	C	\$ (or £, €)
Specific investment cost	$SIC = C/Q_{h,max}$	\$/kWh (or £/kWh, €/kWh)
Lifetime	T_{life}	years (or cycles)

as the energy-storage density, q, this ratio provides a measurement of the minimum space required to store a certain amount of energy:

$$q = \frac{Q_{st}}{V}. \tag{8.1}$$

High storage density is crucial in space-limited applications such as domestic hot water storage. Yet, a higher density limits the maximum heat-transfer area between the heat source/stream (or heat sink) and the storage media, which incurs additional constraints.

The storage temperature range guides the choice of storage media and often rules out a large number of candidates. Non-pressurised water storage is for example limited to storage below 100 °C, while PCMs exhibit poor storage capacities away from their phase-change saturation temperature (due to relatively low specific-heat capacities in the fluid and solid domains).

The storage period varies significantly whether a TES system is designed to limit the performance drop of an ORC engine due to high-frequency time-varying heat-source conditions or to collect and store heat in the summer for providing building heating in the winter. The intended storage duration in TES systems widely varies between these two extreme examples (very short-term intermittency mitigation and inter-seasonal storage). The thermal inertia of a material, also referred to as thermal effusivity, e_{th}, evaluates the pace at which a body approaches its surrounding temperature:

$$e_{th} = \sqrt{\rho c_p k}, \tag{8.2}$$

where ρ is the media density, c_p its specific-heat capacity and k its conductivity. Selecting a material with adequate thermal inertia is key for an effective thermal storage. Together with the thermal-insulation equivalent conductance, G_{ins}, the thermal effusivity provides a useful metric to quantify the heat loss during long-lasting storage. For inter-seasonal storage for example, underground TES (e.g. using boreholes) makes use of the earth high thermal inertia to store heat over a few months. In other words, as earth ground slowly cools down, this medium is very suitable for long-term storage.

The storage and deliverable power corresponds to the heat-transfer rate from the heat source to the heat-storage media during the charging phase and from that media to the heat sink during the discharge, which not only depends on the material thermo-physical and transport properties, but also on the heat-transfer area. As noted by Abhat (1983), a TES system is made of three components: (i) a heat storage substance, (ii) a containment, and (iii) a heat-exchange surface. Fundamental understanding of both heat exchangers and heat-storage media are thus key to the design of an efficient heat battery.

The thermodynamic performance of TES devices is measured by two indicators, respectively derived from the first and second laws of thermodynamics. Firstly, the storage efficiency, η_{st}, is defined as the ratio of the heat recovered in the heat sink, Q_c, to the heat input from the heat source, Q_h:

$$\eta_{st} = \frac{Q_c}{Q_h}. \tag{8.3}$$

The thermal energy input and output terms are obtained from time integrals:

$$Q_h = \int_{\Delta t_h} \dot{m}_h \, (h_{hi} - h_{ho}) \, dt, \tag{8.4}$$

$$Q_c = \int_{\Delta t_c} \dot{m}_c \, (h_{co} - h_{ci}) \, dt, \tag{8.5}$$

where h denotes the specific enthalpy. The first law performance indicator, η_{st}, measures the efficiency with which a TES system stores and delivers back a given amount of thermal energy. It characterises the quantity of energy lost through the overall charge–storage–discharge cycle, but does not account for the decrease in the heat quality. During the successive heat-transfer processes, heat is indeed degraded: the temperature at which heat is recovered by the heat sink is necessarily lower than that at which it was supplied by the heat source.

The inherent irreversibility of heat transfer entails a 'quality' loss that can be quantified via an exergy balance:

$$X_{dest} = X_h - X_c, \tag{8.6}$$

where X_h and X_c are the exergy input and output, which are functions of: (i) the inlet/outlet specific enthalpy and entropy; and (ii) of the dead-state temperature, T_0:

$$X_h = \int_{\Delta t_h} \dot{m}_h \left[(h_{hi} - h_{ho}) - T_0 (s_{hi} - s_{ho}) \right] dt, \qquad (8.7)$$

$$X_c = \int_{\Delta t_c} \dot{m}_c \left[(h_{co} - h_{ci}) - T_0 (s_{co} - s_{ci}) \right] dt. \qquad (8.8)$$

The total exergy destroyed and lost through the overall cycle, X_{dest}, is a direct measure of the energy degradation. The exergetic storage efficency, $\eta_{exg,st}$, is then defined as the ratio of the exergy recovered, X_c, to the exergy input, X_h:

$$\eta_{exg,st} = \frac{X_c}{X_h}. \qquad (8.9)$$

The durability and cost are finally decisive parameters to be considered for the development of suitable TES devices. The specific investment cost (SIC) of a TES system can be defined either as the ratio of the capital expenditure (CAPEX) (i.e. the system manufacturing and installation costs, which correspond to the actual price to be paid to acquire a given asset) to the plant storage capacity (expressed in $/kWh) or as the ratio of the CAPEX to the charging/discharging rate (expressed in $/kW).

8.1.2 Classification of Heat-Storage Media

Heat storage media shall be classified in three main categories:

- sensible heat-storage (SHS) media;
- latent heat-storage media or PCMs;
- (thermo-)Chemical materials, in which energy is stored and released through breaking and reforming molecular bonds in reversible endo-/exothermic reactions.

Figure 8.2 presents a classification of the large variety of non-chemical heat storage media (i.e. single-phase and PCMs) based on their phase and species composition, as first proposed by Abhat (1983) and later completed in several review papers, including the most recent ones from Zalba et al. (2003), Sharma et al. (2009), Zhou, Zhao and Tian (2012) and Sarbu and Sebarchievici (2018).

Sensible heat storage is the simplest and most economical TES strategy whereby the temperature of a given mass of liquid (e.g. water) or solid (e.g. building fabrics, sand or rocks) rises and falls to store and release energy, respectively (see Figure 8.3a). On the other hand, energy stored in PCMs is stored both as sensible heat and latent heat, as schematically presented in Figure 8.3b. Desired properties of SHS materials and PCMs are presented in Sections 8.2 and 8.3, respectively.

While they offer the highest energy storage density due to large endothermic heat of reaction, chemical energy storage systems are still in research and development stage. Advantages and drawback of C-TES solutions are discussed in Section 8.4.

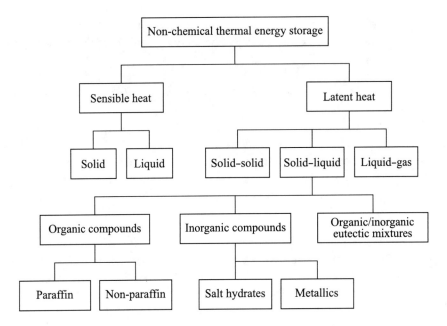

Figure 8.2 Classification of non-chemical TES materials based on phase and species

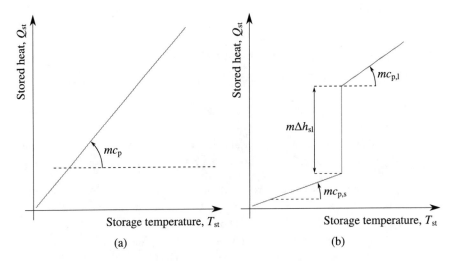

Figure 8.3 Thermal energy storage principle using: (a) SHS media; (b) solid- liquid PCMs. The symbol m represents the storage-substance mass, c_p the specific-heat capacity and Δh_{sl} the latent heat of fusion

8.2   Sensible Heat Storage

The amount of energy stored in a SHS system is a function of the temperature rise of the storage material (from T_0 to T_{st}) and its mass m and specific-heat capacity c_p:

$$Q_{st} = m \int_{h_0}^{h_{st}} dh = m \int_{T_0}^{T_{st}} c_p dT, \tag{8.10}$$

which simplifies if the heat capacity is assumed constant through the charging phase:

$$Q_{st} = m c_p (T_{st} - T_0). \tag{8.11}$$

Similarly, the amount of exergy stored in a fixed-mass SHS system is obtained as:

$$X_{st} = m \int_{T_0}^{T_{st}} c_p \left(1 - \frac{T_0}{T}\right) dT = m c_p \left[(T_{st} - T_0) - T_0 \ln \left(\frac{T_{st}}{T_0}\right)\right]. \tag{8.12}$$

These extensive properties are useful to characterise a storage plant, with a defined mass and operating temperature range, but are inconvenient to specifically compare different storage media. While the storage capacity and hence the storage density depend on the storage process, the volumetric heat capacity offers a material-specific indicator to determine the heat-storage potential of a given medium. This intensive property of the storage substance, defined as the product of the density ρ and specific-heat capacity c_p, is reported in Table 8.2 for a range of liquid- and solid-state SHS media, along with useful thermo-physical properties.

In particular, a key thermal property of a SHS medium is the thermal diffusivity. While the thermal effusivity, or heat-penetration coefficient, characterises the heat transfer at the boundary of the medium (i.e. the pace at which heat is transferred to/from its surface), the thermal diffusivity, $\mathcal{D}$, characterises the heat conduction through the material (i.e. the rate at which the material tends to thermal equilibrium):

$$\mathcal{D} = \frac{k}{\rho c_p}. \tag{8.13}$$

A high thermal diffusivity favours a low temperature difference between the bulk SHS material and the heat source (or heat sink) during the charge (or discharge), which leads to lower exergy destruction through heat transfer.

Suitable sensible TES storage media differ depending on whether high- or low-grade heat is to be stored for heating or cooling purposes.

For low-grade thermal energy storage, both for heating and cooling purposes, water is an ideal storage material and the best liquid medium presented in Table 8.2 due to its high volumetric heat capacity and very low cost, while being limited to storage temperatures above 0 °C (for cold storage) and up to 100 °C. Despite this restriction, water-tank storage is the most common type of SHS system, with applications to domestic hot water, space heating, district heating and TES (Basecq et al. 2013). In the building sector, walls and floor fabrics can also be considered as SHS media. In particular, underfloor heating systems use a concrete layer as a thermal buffer to regulate the room temperature, thus reducing the temperature variation amplitude compared to radiative heaters.

Table 8.2 Thermal properties of liquid- and solid-state sensible heat-storage media

Material	Type	Temperature T_{st} (°C)	Heat Capacity ρc_p (kJ/m^3/K)	Effusivity $\sqrt{k\rho c_p}$ (Ws$^{\frac{1}{2}}$/m^2/K)	Diffusivity $k/(\rho c_p)$ (mm^2/s)
Liquid-state storage media					
Water	–	0–100	4,180[1]	1,588[1]	0.14[1]
Ethanol	Organic	(−114) to 78	1,891[1]	558[1]	0.08[1]
Propane	Organic	(−188) to (−42)	1,304[2]	413[2]	0.1[2]
Therminol 55	HTF[3]	(−28) to 290	1,665[1]	462[1]	0.08[1]
Therminol 66	HTF[3]	(−3) to 345	1,572[1]	430[1]	0.08[1]
HITEC[4]	Molten salt	222 to 580	2,621[1]	–	–
Solid-state storage media[5]					
Sand	Mineral	<1,500	1,060–1,500	400–580	0.11–0.25
Concrete	Mineral	<1,800	1,970–2,120	990–1,890	0.24–0.86
Granite	Mineral	<1,215	2,050–2,140	2,020–2,730	0.93–1.7
Aluminium	Metal	<660	2,425	22,800	88.6
Cast iron	Metal	<1,150	6,612	21,804	10.9

[1] Thermodynamic properties at 20 °C.

[2] Thermodynamic properties at −45 °C.

[3] Therminol heat transfer fluids (HTFs) are modified terphenyls (hydrocarbons). Detailed information are available for the Therminol 55 and 66 at www.therminol.com/sites/therminol/files/documents/TF-25_Therminol_55.pdf and www.therminol.com/sites/therminol/files/documents/TF-8695_Therminol-66_Technical_Bulletin.pdf, respectively.

[4] Thermal properties of the HITEC molten salt (60% NaNO$_3$ + 40% KNO$_3$) were determined experimentally by Fernández et al. (2015).

[5] Thermal properties of minerals vary according to their porosity and composition.

For high-grade cold-storage applications (typically below −20 °C), organic fluids, including alcohols and hydrocarbons, are suitable liquid-state candidates, along with solid-state materials.

For medium- to high-grade hot-storage applications (i.e. for storage temperatures ranging from 100 to 300 °C and above 300 °C, respectively), solid-state substances are preferred, while specifically engineered (such as the Therminol) heat-transfer fluids and molten salts offer viable liquid-state storage solutions. As outlined by Velraj (2016), molten salts show a number of advantages besides their high melting temperature, including high thermal stability, non-flammability and a low vapour pressure, which removes the need for pressurised vessels.

As reported in the technology brief by the International Renewable Energy Agency (2013), SHS systems are the most economical in comparison with latent and chemical TES solutions. The SIC of a complete sensible TES device ranges from 0.1 to 10 €/kWh.

Table 8.3 Thermodynamic properties of solid–solid PCMs

Material	Crystallisation Temperature (°C)	Latent Heat of Crystallisation (kJ/mol)	Fusion Temperature (°C)	Latent Heat of Fusion (kJ/mol)
Pentaerythritol	188	41.26	260	5.02
Pentaglycerine	83	23.12	198	5.43
Neopentylglycol	41	13.63	126	4.6

8.3 Latent Heat Storage

Latent heat storage is particularly attractive from a thermodynamic perspective due to two main intrinsic advantages of PCMs, namely: their high energy-storage density and their ability to store energy at a constant temperature. Energy storage using phase transitions between liquid, solid and gas states are discussed hereafter.

8.3.1 Liquid- and Solid–Gas Latent Heat Storage

Even though latent heat of vaporisation/condensation and sublimation/deposition are particularly high, heat storage using liquid–gas or solid–gas phase transitions are in practice not suitable due to the large specific volume differences between the high- and low-enthalpy states. In order to conserve their high energy-storage density potential, pressurised vessels should be employed, thus incurring high additional costs.

8.3.2 Solid–Solid Latent Heat Storage

Solid–solid PCMs rely on the heat of crystallisation to store/release energy through the transitions to/from a mesophase. For example, pentaerythritol experiences a solid–solid phase transition between 187 and 189 °C, whereby the tetragonal crystalline structure of the material changes into a disordered face-centred cubic crystalline structure. The latent heat of crystallisation of such a transition was measured experimentally by Venkitaraj and Suresh (2017), who found that the transformation was accompanied by an enthalpy change of 248 kJ/kg (233 kJ/kg after 100 cycles).

Polyalcohols such as pentaerythritol ($C_5H_{12}O_4$), pentaglycerine ($C_5H_{12}O_3$) and neo-pentylglycol ($C_5H_{12}O_2$) exhibit solid–solid latent heat comparable to the latent heat of fusion of various paraffins, which are typical solid–liquid PCMs. Their thermodynamic properties and solid–solid transition temperature are presented in Table 8.3. For an exhaustive list of solid–solid PCMs, please refer to the complete review proposed by Fallahi et al. (2017).

The main advantage of solid–solid PCMs is the small volume change undergone during the phase transition, while their main drawbacks are a low thermal conductivity and the deterioration of their structure after repeated charge/discharge cycles.

8.3.3   Solid-Liquid Latent Heat Storage

The amount of energy stored in a solid–liquid PCM is a function of the temperature rise of the storage material (from T_0 to $T_{st} > T_m$, the melting temperature) and its mass m and thermo-physical properties:

$$Q_{st} = m \int_{h_0}^{h_{st}} dh = m \left[\int_{T_0}^{T_m} c_{p,s} dT + \Delta h_{sl} + \int_{T_m}^{T_{st}} c_{p,l} dT \right], \qquad (8.14)$$

where Δh_{sl} is the latent heat of fusion, and $c_{p,s}$ and $c_{p,l}$ the specific heat capacities of the solid and liquid PCM, respectively. Assuming the latter are constant through the charging phase, the expression for the energy stored in solid–liquid PCMs simplifies to:

$$Q_{st} = m \left[c_{p,s} (T_m - T_0) + \Delta h_{sl} + c_{p,l} (T_{st} - T_m) \right]. \qquad (8.15)$$

Similarly, the amount of exergy stored in a PCM can be obtained as:

$$X_{st} = m \int_{T_0}^{T_{st}} \left(1 - \frac{T_0}{T} \right) \delta Q$$

$$= m\, c_{p,s} \left[(T_m - T_0) - T_0 \ln \left(\frac{T_m}{T_0} \right) \right] + m\, \Delta h_{sl} \left(1 - \frac{T_0}{T_m} \right)$$

$$+ m\, c_{p,l} \left[(T_{st} - T_m) - T_0 \ln \left(\frac{T_{st}}{T_m} \right) \right]. \qquad (8.16)$$

A large variety of PCMs are available, including organic, inorganic and eutectic compounds. Table 8.4 provides a non-exhaustive list of promising PCMs, classified by nature and melting temperature.

Abhat (1983) proposes a list of desired (or ideal) properties of solid–liquid PCMs, which can be classified under four main categories, namely: thermodynamic, kinetic, chemical and economic properties.

Thermodynamic Properties

- melting temperature in desired range;
- high latent heat of fusion per unit volume;
- high thermal conductivity, specific heat and density;
- small specific volume change on phase transition;
- small vapour pressure at operating temperatures;
- congruent melting.

Kinetic Properties

- high nucleation rate to avoid supercooling;
- high crystal-growth rate to meet heating/cooling demands.

Chemical Properties

- fully reversible freezing/melting cycle;

Table 8.4 Thermal properties of selected liquid-solid PCMs.

PCM	Melting Temperature (°C)	Latent Heat of Fusion (kJ/kg)	Thermal Conductivity (W/m/K)
Paraffin (organic compounds)			
Paraffin C_{13}-C_{24}	22–24	189	0.21
Paraffin C_{18}	28	244	0.15
Paraffin C_{20}	36.7	246	–
Paraffin C_{30}	65.4	251	–
Paraffin C_{34}	75.9	269	–
Non-Paraffin organic compounds			
Polyglycol E600	22	127.2	0.19
D-Lattic acid	26	184	–
Heptaudecanoic acid	60.6	189	–
Bee wax	61.8	177	–
Quinone	115	171	–
Succinic anhydride	119	204	–
Benzamide	127.2	169.4	–
Salt hydrates			
$CaCl_2 \cdot 12H_2O$	29.8	174	–
$LiNO_3 \cdot 3H_2O$	30	189	–
$CoSO_4 \cdot 7H_2O$	40.7	170	–
$Ca(NO_3) \cdot 4H_2O$	47	153	–
$NaOH \cdot H_2O$	64.3	273	–
$MgCL_2 \cdot 6H_2O$	117	167	
Metallic inorganic compounds			
Mercury	−38.87	11.4	8.34
Gallium	30	80.3	–
Cerrolow eutectic	58	90.9	–
Sodium	58	90.9	86.9
Cerrobend eutectic	70	32.6	–
Lithium	186	433.78	41.3
Bismuth	271.4	53.3	8.1

Data gathered from several review papers: (Ge et al. 2013); (Sarbu and Sebarchievici 2018); (Abhat 1983);(Sharma et al. 2009); (Zhou, Zhao and Tian 2012); and (Zalba et al. 2003)

- chemically stable stability;
- no degradation after a large number of freezing/melting cycles;
- no corrosiveness, toxicity, flammability.

Economic Properties

- low SIC;
- large-scale availability.

   Regarding the latter category, the cost of LHS systems is typically higher than that of a SHS plant, as substantial costs are associated with the heat exchanger system. Indeed, the stream-to-PCM heat transfer must be sufficient to achieve the high charging rates that characterise PCM storage devices. According to the technology brief proposed by the International Renewable Energy Agency (2013), the cost of a PCM-based TES system ranges between 10 and 50 €/kWh. Yet, it must be kept in mind that the SIC is not the only techno-economic performance indicator. It does not account for the charging rate nor for the overall exergetic efficiency.

8.3.4   PCM Containment

The high heat-storage density and relatively low thermal conductivity of PCMs pose a great challenge in designing the container/PCM/heat-exchanger assembly. Due to the heat-diffusion thermal resistance within the material, large heat-exchange areas are required between the PCM and the heat-transfer fluid to provide satisfactory charging/discharging rates, which is made difficult by the reduced heat-storage volume. Three main strategies are used for PCM containment, namely: bulk storage, encapsulation and shape stabilisation.

PCM bulk storage systems do not offer large heat-transfer area and thus rely on heat-transfer enhancement techniques such as insertion of fins, nanoparticles addition to increase the effective thermal conductivity or use of direct-contact heat exchangers.

Macro- and micro-encapsulation aim at optimising the packing surface area per unit volume. Macro-encapsulation consists in filling the PCM within cylinders, spheres or slabs, past which the heat-transfer fluid circulates. An example of macro-encapsulation in a cylinder is proposed by Höhlein, König-Haagen and Brüggemann (2018). On the other hand, micro-encapsulation refers to the development of micron-scale capsules (typically micron-scale spheres), made of a polymer shell containing a PCM core. The obtained powder can be used either as is for storing heat through thermal diffusion or be mixed with a fluid (e.g. water) to obtain a slurry that can be circulated.

Shape-stabilised PCM plates, spheres or cylinders are obtained by combining the molten PCM with a support material (e.g. high-density polyethylene or styrene-butadiene-styrene) heated up above its glass transition temperature.

8.4   Chemical Thermal Energy Storage

Chemical thermal energy storage relies on the use of reversible endo- and exothermic reactions to store thermal energy. The inter-molecular bonds between chosen reactants can be broken through heat addition, thus creating a group of reactive components that can be stored separately, see Table 8.5. The heat supplied during the charging process can then be recovered by re-combining the reactive components. C-TES systems are distinguished according to the type of chemical process involved, which is either a thermo-chemical (sorption) process or a chemical reaction.

Thermo-chemical TES systems store thermal energy through endothermic desorption, the process during which the chemical bonds that retain water (called the sorbate) within a sorbent material are broken. The heat is then recovered through

Table 8.5 Chemically reacting substances and sorption material pairs for heat-storage applications

Material		Storage Density
Chemical reactions		
Magnesium sulphate	$MgSO_4 \cdot 7H_2O \Leftrightarrow MgSO_4 + 7H_2O$	2.8 GJ/m^3 [1]
Iron carbonate	$FeCO_3 \Leftrightarrow FeO + CO_2$	2.6 GJ/m^3 [1]
Iron hydroxide	$Fe(OH)_2 \Leftrightarrow FeO + H_2O$	2.2 GJ/m^3 [1]
Absorption		
Ammonia-water	NH_3-H_2O	1.3 GJ/kg [2]
Water-lithium bromide	H_2O-LiBr	2.0 GJ/kg [3]
Water-sodium hydroxide	H_2O-NaOH	1.6 GJ/kg [4]
Adsorption		
Water-zeolite 4A	H_2O-zeolite	0.08 GJ/kg [1]
Water-silica gel	H_2O-silica	0.14 GJ/kg [1]

[1] Provided by Pinel et al. (2011).

[2] Provided by Hui et al. (2011) for 90% mass fraction of absorbent after desorption.

[3] Provided by Hui et al. (2011) for a mass fraction of absorbent after desorption of 58.8%.

[4] Provided by Hui et al. (2011) for a mass fraction of absorbent after desorption of 33.5%.

the exothermic sorption process, during which the bonds linking sorbate and sorbent components are restored. Sorption processes involve both ab- and ad-sorption mechanisms, depending on whether these reactions occur in the bulk (i.e. molecules being absorbed within the volume) or at the surface of a material. Adsorption materials – to the surface of which molecules adhere – are usually solid, while absorption media are usually liquid. Adsorption and absorption are thus commonly referred to as liquid- and solid-sorption.

Chemical reaction-based TES systems involve the separation and recombination of reacting substances. As shown in Figure 8.4, this storing strategy is mainly attractive due to its very high energy-storage density. Yet, a number of obstacles hinder their development, including the degradation of reactants after repeated cycles and corrosion issues. In addition, the development of cost-effective chemical reaction-based TES systems requires acceptable charging and discharging rates, hence large specific surface area or highly enhanced heat transfer. The reduced size of reaction-based C-TES systems turns out to be both their main advantage and main drawback, as it strongly limits the available area to drive the charge and discharge heat transfer.

8.5   Thermodynamic Modelling

The heat transfer between the heat-source (or heat-sink) stream and the heat-storage material involves: (i) forced convection on the heat-transfer fluid side, and (ii) heat

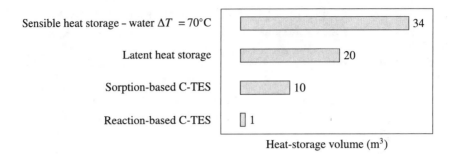

Figure 8.4 Comparison of TES systems energy-storage density: estimate of the required volume of sensible, latent and chemical media to store 6.7 MJ of heat. Numbers presented here are informative only and based on a conservative analysis performed by Hadorn (2008)

diffusion-advection within the storage medium. Heat-transfer correlations are presented in detail in Section 8.5.1 to evaluate the former. As for the latter, the transient heat propagation within single-phase stagnant liquid and solid storage materials is described in Section 8.5.2, while Section 8.5.3 presents analytical and numerical methods to predict the instantaneous heat transfer in PCMs.

8.5.1   Forced Convective Heat Transfer on the Heat-Transfer Fluid Side

Whether the heat-storage medium is a single-phase liquid or solid material, or a solid–liquid PCM, two main arrangements are used to transfer heat between the heat-storage medium and the heat-transfer fluid (i.e. either the heat-source or heat-sink stream). The latter either flows within a coiled pipe or channels *embedded within the heat-storage medium* (as shown in Figure 8.5a) or flows *around the storage material*, for example arranged in a packed bed or tube bundle (see Figure 8.5b). The latter configuration is particularly well suited for gas-to-solid heat transfer or for macro-encapsulated PCMs.

Forced convection heat-transfer rates depend on both the fluid thermo-physical properties and the flow configuration, as the geometry largely influences the fluid-to-wall thermal boundary layer. A number of correlations are presented next to predict the convective contribution to the overall heat transfer, while the heat diffusion within the storage media will be addressed in Sections 8.5.2 and 8.5.3, for sensible and LHS, respectively.

Forced Convection in a Pipe

Sieder and Tate (1936) developed a heat-transfer correlation for a laminar flow in a pipe (i.e. for Re $< 2{,}000$) accounting for the dependency of the fluid viscosity upon the temperature:

$$\mathrm{Nu} = 1.86\,\mathrm{Re}^{1/3}\,\mathrm{Pr}^{1/3}\left(\frac{L}{D}\right)^{-1/3}\left(\frac{\mu_b}{\mu_w}\right)^{0.14}, \qquad (8.17)$$

where D and L are the internal pipe diameter and length, respectively, and μ_w the fluid dynamic viscosity at wall temperature. Nu, Re and Pr are the Nusselt Reynolds and Prandtl number, respectively, defined as:

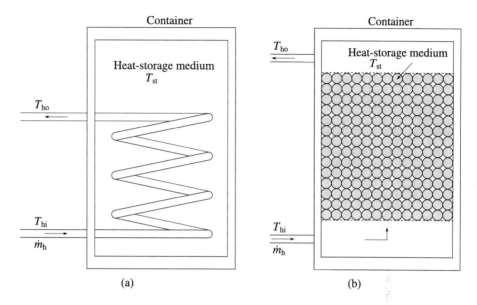

Figure 8.5 Schematics of typical arrangements in TES systems: (a) heat-transfer fluid flowing in a coiled pipe embedded within the heat-storage material (bulk storage); (b) heat-transfer fluid flowing through a packed bed of spherical particles or pebbles, which could be either plain balls of SHS material or encapsulated PCM

$$\text{Nu} = \frac{\alpha D}{\lambda_b}, \quad \text{Re} = \frac{v_b D}{v_b}, \quad \text{and} \quad \text{Pr} = \frac{c_{p,b}\,\mu_b}{\lambda_b}, \tag{8.18}$$

where α is the heat-transfer coefficient. The terms v_b, $c_{p,b}$, λ_b v_b, μ_b refer to the velocity, specific-heat capacity, thermal conductivity and kinematic and dynamic viscosities of the bulk fluid, respectively. Bulk fluid properties (denoted with the subscript b) are obtained at the so-called mean or average bulk fluid temperature, T_b, defined for a flow entering the pipe at T_i and leaving at T_o as:

$$T_b = \frac{T_i + T_o}{2}. \tag{8.19}$$

The convective heat-transfer coefficient for a turbulent in-pipe flow can be obtained using the correlation developed by Gnielinski (1975):

$$\text{Nu} = \frac{f^* \,(\text{Re} - 1{,}000)\,\text{Pr}}{1 + 12.7\,\sqrt{f^*}\,\left(\text{Pr}^{2/3} - 1\right)}, \tag{8.20}$$

for Prandtl numbers between 0.5 and 2,000 and Reynolds numbers between 3,000 and $5 \cdot 10^6$. The modified Darcy friction factor f^* can be obtained from the Petukhov (1970) correlation for smooth tubes:

$$f^* = \frac{1}{8}\,(0.79 \log\,(\text{Re}) - 1.64)^{-2}. \tag{8.21}$$

Alternatively, Dittus and Boelter (1930) proposed a correlation for turbulent flows in pipes valid for $0.6 \leqslant \text{Pr} \leqslant 160$ and $\text{Re} \leqslant 10{,}000$:

$$\text{Nu} = 0.023 \, \text{Re}^{4/5} \, \text{Pr}^{n}, \tag{8.22}$$

with $n = 0.3$ if the fluid is being cooled and $n = 0.4$ if the fluid is being heated.

Forced Convection on a Flat Plate

For forced convection on a flat plate, Whitaker (1972) recommends using the following correlations:

$$\text{Nu}_x = 0.332 \, \text{Re}_x^{1/2} \, \text{Pr}^{1/3} \quad \text{for a laminar boundary layer;} \tag{8.23}$$

$$\text{Nu}_x = 0.029 \, \text{Re}_x^{0.8} \, \text{Pr}^{0.43} \quad \text{for a turbulent boundary layer.} \tag{8.24}$$

In Equations (8.23) and (8.24), Nu_x and Re_x refer to the local Nusselt and Reynolds numbers, at a distance $x \in [0\,;L]$ downstream of the plate edge, with L being the plate length (in the flow direction).

Based on the previous equations and assuming that the laminar-to-turbulent transition occurs at a Reynolds of $2 \cdot 10^5$, Whitaker (1972) derives an average Nusselt number for the plate that equals:

$$\text{Nu} = 0.036 \left(\text{Re}^{0.8} \, \text{Pr}^{0.43} - 17{,}400 \right) + 297 \, \text{Pr}^{1/3}, \tag{8.25}$$

with non-dimensional numbers defined according to the plate length, L:

$$\text{Nu} = \frac{\alpha L}{\lambda_b} \quad \text{and} \quad \text{Re} = \frac{v_b L}{\nu_b}. \tag{8.26}$$

Flow Through Packed Beds

A packed bed is a porous medium characterised by a porosity (or void fraction), φ, which is the ratio of the void volume in the bed (i.e. the volume occupied by the fluid) to the total volume of the bed. The forced convective fluid-to-bed heat transfer can be described either through the definition of a log-mean film heat-transfer coefficient or by defining a local volumetric heat-transfer coefficient between the solid and fluid phases.

The log-mean heat-transfer coefficient, α_{lm}, is defined from an overall energy balance:

$$\dot{Q}_{lm} = \alpha_{lm} a_s V \Delta T_{lm}, \tag{8.27}$$

where $\dot{Q}_{lm}$ is the total fluid-to-solid heat-transfer rate, V_{bed} is the total packed-bed volume and ΔT_{lm} the log-mean temperature difference across the bed. The packing surface area per unit volume, a_s, is a characteristic of the porous medium, defined as:

$$a_s = (1 - \varphi) \frac{A_p}{V_p}, \tag{8.28}$$

where A_p and V_p are the area and volume of a single solid particle forming the packed bed. Whitaker (1972) analysed a large number of experimental data and derived a useful and simple correlation for light gases flowing through packed beds:

$$\text{Nu}_{\text{lm}} = \text{Pr}^{1/3} \left(0.5 \, \text{Re}_{\text{lm}}^{1/2} + 0.2 \, \text{Re}_{\text{lm}}^{2/3} \right). \tag{8.29}$$

The specific log-mean Nusselt, Nu_{lm}, and Reynolds, Re_{lm}, numbers are defined as:

$$\text{Nu}_{\text{lm}} = \frac{\alpha \, D_{\text{p}}}{\lambda_{\text{f}}} \frac{\varphi}{1 - \varphi}; \quad \text{and} \quad \text{Re}_{\text{lm}} = \frac{\overline{v_{\text{f}}} \, D_{\text{p}}}{v_{\text{f}} \, (1 - \varphi)}, \tag{8.30}$$

where D_{p} is the particle diameter (i.e. the mean diameter of the pebbles, grains or spheres forming the bed), while $\overline{v_{\text{f}}}$ denotes the fluid superficial velocity, which depends on the mass flow rate, $\dot{m}_{\text{f}}$, fluid density, ρ_{f}, and the cross-sectional area of the bed, A:

$$\overline{v_{\text{f}}} = \frac{\dot{m}_{\text{f}}}{\rho_{\text{f}} \, A}. \tag{8.31}$$

The volumetric heat-transfer coefficient, α_{fs}, describes the local heat transfer from the fluid phase (subscript f) to the solid one (subscript s). Due to the local thermal non-equilibrium, macroscopic energy conservation equations are written for both phases:

$$(1 - \varphi)\rho_{\text{s}} c_{\text{p,s}} \frac{\partial \overline{T_{\text{s}}}}{\partial t} = \alpha_{\text{fs}} \left(\overline{T_{\text{f}}} - \overline{T_{\text{s}}} \right) + G_{\text{s}} \nabla \overline{T_{\text{s}}}, \qquad \text{for the solid phase;} \tag{8.32}$$

$$\varphi \rho_{\text{f}} c_{\text{p,f}} \frac{\partial \overline{T_{\text{f}}}}{\partial t} + \rho_{\text{f}} c_{\text{p,f}} \overline{v_{\text{f}}} \nabla \overline{T_{\text{f}}} = \alpha_{\text{fs}} \left(\overline{T_{\text{s}}} - \overline{T_{\text{f}}} \right), \qquad \text{for the fluid phase.} \tag{8.33}$$

Please note that the temperatures $\overline{T_{\text{f}}}$ and $\overline{T_{\text{s}}}$ used in the macroscopic representation in Equations (8.32) and (8.33) refer to local volume-averaged temperatures. G_{s} represents the equivalent thermal conductivity through the solid matrix.

The local fluid-to-solid heat-transfer volumetric heat-transfer coefficient, α_{fs}, accounts for both the convective thermal resistance (i.e. the fluid-to-pebble surface resistance) and the diffusive thermal resistance (i.e. the resistance induced by the heat conduction within the solid pebbles). For poorly conductive solid materials, the gradients within the pebbles must be accounted for to describe the full fluid-to-solid heat transfer. On the contrary, if the solid particle Biot number is significantly smaller than unity, the fluid-to-solid thermal resistance is dominated by the convective contribution and the local heat-transfer coefficient can be estimated from adequate convection correlations.

Löf and Hawley (1948) proposed a simple correlation to predict the convective air-to-particles volumetric heat-transfer coefficient in packed beds:

$$\alpha_{\text{fs}} = 650 \left(\frac{\dot{m}_{\text{f}}}{A \, D_{\text{p}}} \right)^{0.7}. \tag{8.34}$$

8.5.2   Dynamic Modelling of Sensible Heat Storage

The heat diffusion within single-phase solid or stagnant liquid materials is a multi-dimensional transient problem with often time-dependent non-homogeneous boundary conditions. The resolution of the full three-dimensional conjugate heat-transfer problem requires the use of computationally expensive CFD methods, while lumped-capacity and reduced-order solutions are easily obtained without scarifying accuracy, provided that adequate assumptions are used.

The heat transfer from/to the flowing heat-transfer fluid occurs through a heat-exchange surface, S. From the first law of thermodynamics, one can thus derive the following differential equation to describe the temporal evolution of the heat-storage material temperature T_{st} during the charging/discharging processes:

$$mc_p \frac{d \langle T_{st} \rangle}{dt} = \alpha S (T_b - \langle T_{st} \rangle), \qquad (8.35)$$

where T_b is the mean bulk temperature in the heat-source or heat-sink streams and $\langle T_{st} \rangle$ the mass-averaged temperature of the heat-storage medium. The mean heat-transfer coefficient, α, accounts for both the convective (between the surface and the external flow) and conductive (within the storage body) contributions, namely: α_{cv} and α_{cd}. The competition between those two modes of heat transfer shall decide whether a refined analysis of the temperature distribution within the storage material is required.

The Biot number, Bi, is a simple index used in heat-transfer analysis to compare the magnitude of both contributions:

$$\text{Bi} = \frac{\alpha_{cv} D}{\lambda}, \qquad (8.36)$$

where D is a characteristic dimension of the heated/cooled body and λ/D is an estimate of the equivalent conductive heat-transfer coefficient.

If the heat diffusion within the body offers little resistance to the overall heat transfer, that is, if $\text{Bi} \ll 1$, then the surface convection is the dominant heat-transfer mode and α ($= \alpha_{cv}$) can be directly obtained from one of the correlations listed in Section 8.5.1. In addition, as the surface heat-transfer rate is slow in comparison with the rate at which the material tends to thermal equilibrium, one can assume that the storage medium temperature is homogeneous through the heating/cooling process:

$$\langle T_{st} \rangle (t) = \frac{1}{m} \int_V \rho \, T_{st}(\mathbf{x}, t) \, d\mathbf{x} \rightarrow T_{st}(t). \qquad (8.37)$$

The heated/cooled body is thus treated as lump and the charging/discharging processes can be fully described by lumped-capacity transient thermal analyses.

Lumped-Capacity Transient Solution
The lumped-capacity solution to a conjugate heat-transfer problem is a simple and very useful tool, provided that the Biot number is small, that is, provided that the thermal resistance due to heat diffusion within the material is small compare to that of the surface convective heat transfer.

Assuming a constant bulk flow temperature, that is, for $dT_b/dt = 0$, the lumped-capacity solution to Equation (8.35) is given as:

$$\frac{T_{st} - T_b}{T_0 - T_b} = e^{-\frac{t}{\tau}} \qquad \text{with} \qquad \tau = \frac{\alpha S}{m\,c_p}, \tag{8.38}$$

where T_0 is the initial storage-material temperature and τ the heat-transfer time constant.

For an arbitrarily time-varying bulk temperature $T_b(t)$, the lumped-capacity temperature evolution is given by:

$$T_{st} = T_b + (T_0 - T_b(t=0))\,e^{-\frac{t}{\tau}} - e^{-\frac{t}{\tau}} \int_0^t e^{\frac{s}{\tau}}\,\frac{dT_b(s)}{ds}\,ds. \tag{8.39}$$

Packed-Bed Sensible Energy Storage

The transient analysis of a hot or cold gas flowing through a packed bed of nearly spherical solid particles is a problem often encountered in both latent and sensible heat TES systems, as the large heat-transfer area in pebble beds and the locally turbulent flows provide an efficient heat exchange. This arrangement is typically used with gas-phase heat-source or heat-sink streams.

For low pebble-specific Biot numbers, the temperature gradients within the pebbles can thus be neglected and, assuming a non-compressible flow and neglecting axial and radial conduction through the solid matrix, the energy conservation equations given by Equations (8.32) and (8.33) simplify to:

$$(1 - \varphi)\rho_s c_{p,s}\frac{\partial T_s}{\partial t} = \alpha_{fs}(T_f - T_s) \qquad \text{for the solid phase;} \tag{8.40}$$

$$\rho_f c_{p,f}\overline{v_f}\frac{\partial T_f}{\partial x} = \alpha_{fs}(T_s - T_f) \qquad \text{for the fluid phase.} \tag{8.41}$$

The numerical integration of Equations (8.40) and (8.41) can be efficiently performed using the semi-analytical routine described by White (2011). The fluid and solid temperatures at node i and time-step n are functions of the solution at $(n, i-1)$ and $(n-1, i)$:

$$\begin{pmatrix} 1 & \dfrac{a-1}{2} \\ \dfrac{b-1}{2} & 1 \end{pmatrix}\begin{pmatrix} T_f|_i^n \\ T_s|_i^n \end{pmatrix} = \begin{pmatrix} \dfrac{a-1}{2}T_s|_{i-1}^n + a\,T_f|_{i-1}^n \\ \dfrac{b-1}{2}T_f|_i^{n-1} + b\,T_s|_i^{n-1} \end{pmatrix}, \tag{8.42}$$

where a and b are obtained analytically and expressed as:

$$a = \exp\left(-\frac{\alpha_{fs}}{\rho_f c_{p,f}\overline{v_f}}\Delta x\right) \qquad \text{and} \qquad b = \exp\left(-\frac{\alpha_{fs}}{(1-\varphi)\rho_s c_{p,s}}\Delta t\right), \tag{8.43}$$

with Δx and Δt the mesh spacing and the time-step size, respectively.

The unknown temperatures are easily obtained by inverting the left-hand side matrix in Equation (8.42). Mesh refinement techniques and adjustable time steps can be used with this integration scheme (even though very stable) for highly anisotropic beds and time-varying boundary conditions.

These simple solutions to convective heat transfer in porous media problems provide reliable first estimates of both the charging time and the temperature differences across the packed bed. It must be noted, however, that a number of assumptions were made (mainly: non-compressible gas flow and no radial nor radial heat conduction through the solid matrix). The comprehensive modelling of such a problem requires a detailed analysis of the pore-scale phenomena to derive effective macroscopic properties through up-scaling methods (e.g. volume-averaging techniques), as described by Quintard and Whitaker (1994). Refined analysis of transport phenomena in porous media however falls out of the scope of this book, which aims at providing simple tools to evaluate the thermodynamic performance of heat-conversion and heat-storage technologies.

Example 8.1   For 10-mm diameter spherical concrete pebbles (with $\rho_s = 2240$ kg/m^3, $c_{p,s} = 880$ J/kg/K, $\lambda_s = 0.8$ W/m/K) arranged in a packed bed of porosity $\varphi = 0.3$, past which air is entering at 50 °C at $\overline{v_f} = 5$ cm/s, Equation (8.34) predicts a volumetric fluid-to-solid heat-transfer coefficient $\alpha_{fs} = 2128$ W/m^3/K. Thanks to the relationship between the volumetric and local surface heat-transfer coefficients:

$$\alpha_{cv} = \frac{\alpha_{fs}}{a_s}, \quad \text{where} \quad a_s = (1 - \varphi)\frac{A_p}{V_p}, \tag{8.44}$$

the convective heat-transfer coefficient at the surface of a concrete sphere is found to be $\alpha_{cv} = 5$ W/m^2/K. The solid particle Biot number that compares surface convective heat transfer to the inner-particle diffusion thus equals:

$$\text{Bi}_p = \frac{\alpha_{cv} D_p}{\lambda_s} = \frac{5 \times 10.10^{-3}}{0.8} = 0.06 \ll 1. \tag{8.45}$$

Equations (8.40) and (8.41) can thus be used to predict the time-resolved temperature profiles and the fluid-to-bed instantaneous heat-transfer rate.

For a 2-m long packed bed of concrete spheres with a cross-sectional area of 10 m^2 initially at $T_0 = 20$ °C, heated up by air entering at $T_{hi} = 50$ °C, bed temperature profiles are predicted at different times in Figure 8.6.

Figure 8.6 shows that the heating front (or thermocline) appears to become progressively less steep, yet propagates within the porous bed at a nearly constant velocity, v_{hf}, which can be approximated by an enthalpy balance across the moving thermocline:

$$v_{hf} = \frac{\rho_f c_{p,f} \overline{v_f}}{(1 - \phi)\rho_s c_{p,s} + \rho_f c_{p,f}}. \tag{8.46}$$

This result, which is independent of the temperature difference between the bed and the flowing fluid, is a useful result to quickly estimate the charging/discharging time of a packed pebble bed.

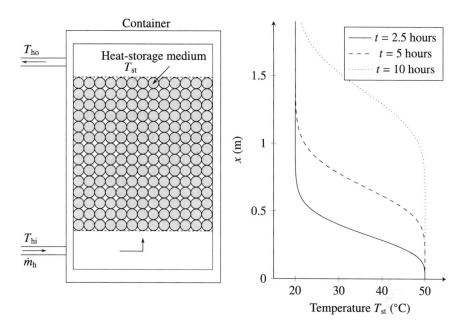

Figure 8.6 Solid temperature profiles plotted at different times during the charging of a 2-m long bed initially at $T_0 = 20\,^\circ\text{C}$, formed of 10-mm diameter spherical concrete pebbles, heated up by air entering at $T_{hi} = 50\,^\circ\text{C}$ at 5 cm/s

Useful One-Dimensional Analytical Solutions

For large Biot numbers, the heat diffusion through the solid or liquid material is too slow to ignore the thermal resistance induced by the internal temperature gradients. In complex geometries, computational fluid dynamics (CFD) tools can provide detailed solutions to the conjugate heat transfer. Yet, a number of one-dimensional analytical solutions can often be used to obtain the exact temperature evolution of the storage medium, or provide a reliable first-estimate of the storage dynamics.

Heat Diffusion in Bounded Elements
Useful theoretical results include the analytical solution to the transient heat conduction problem in a one-dimensional slab, initially at T_0 and subject to a sudden change in surface temperature to T_w:

$$\frac{\partial T}{\partial t} = \mathcal{D}\frac{\partial^2 T}{\partial x^2}; \tag{8.47}$$

$$T(-L, t > 0) = T(L, t > 0) = T_w; \tag{8.48}$$

$$T(x, t = 0) = T_0. \tag{8.49}$$

The general series solution to this problem is provided by Lienhard (2019):

$$\frac{T - T_w}{T_0 - T_w} = \frac{4}{\pi} \sum_{n=odd}^{\infty} \frac{1}{n} \exp\left(-\text{Fo}\left(\frac{n\pi}{2}\right)^2\right) \sin\left(\frac{n\pi}{2}\xi\right), \tag{8.50}$$

Table 8.6 Series coefficients for transient analytical solutions to one-dimensional heat diffusion in slabs, cylinders and spheres subject to convective boundary conditions.

	A_n	K_n	F_n
Slab	$\dfrac{2\sin F_n}{F_n + \sin F_n \cos F_n}$	$\cos\left(\dfrac{F_n x}{L}\right)$	$\cos F_n = \dfrac{F_n}{\mathrm{Bi}_L}$
Cylinder	$\dfrac{2 J_1(F_n)}{F_n\left[J_0{}^2(F_n) + J_1{}^2(F_n)\right]}$	$J_0\left(F_n\dfrac{r}{R}\right)$	$F_n J_1(F_n) = \mathrm{Bi}_R J_0(F_n)$
Sphere	$2\,\dfrac{\sin F_n - F_n\cos F_n}{F_n - \sin F_n\cos F_n}$	$\dfrac{R}{F_n r}\sin\left(\dfrac{F_n r}{R}\right)$	$F_n \cot F_n = 1 - \mathrm{Bi}_R$

J_0 and J_1 are Bessel functions of the first kind

where Fo (Fourier number) and ξ are dimensionless time and distance coordinates:

$$\mathrm{Fo} = \frac{\mathcal{D} t}{L^2}, \quad \text{and} \quad \xi = 1 + \frac{x}{L}. \tag{8.51}$$

If the material initially at T_0 is symmetrically heated (or cooled) by a homogeneous convective surface heat flux from a flowing fluid with a bulk temperature T_b, the boundary conditions of the problem would become:

$$\alpha\,(T_\mathrm{b} - T)_{x=\pm L} = -\lambda\left(\frac{\partial T}{\partial x}\right)_{x=\pm L}. \tag{8.52}$$

Again, Lienhard (2019) provides us with the general solution:

$$\frac{T - T_\mathrm{b}}{T_0 - T_\mathrm{b}} = \sum_{n=1}^{\infty} A_n \exp\left(-K_n{}^2\,\mathrm{Fo}\right) F_n, \tag{8.53}$$

where the values of the A_n, K_n and F_n terms are given in Table 8.6 for slabs ($x \in [-L\,;\,L]$), cylinders ($r \in [0\,;\,R]$) and spheres ($r \in [0\,;\,R]$).

Heat Diffusion in Semi-Infinite Regions
As emphasised by Lienhard (2019), a body exposed to a sudden change in surface temperature is infinitely larger than the region initially influenced by this change. In other words, a body can be considered a semi-infinite body, at least until the thermal boundary layer expands up to another boundary of the body (e.g., the other side of a slab) or up to reaching a singular point such as the centre of a sphere. In many heat-transfer problems, the analytical solution to the semi-infinite heat-diffusion problem thus comes in handy. This problem is described using the classical heat equation (Eq. (8.47)) and the following boundary and initial conditions:

$$T(x = 0, t > 0) = T_\mathrm{w}; \tag{8.54}$$

$$T(x, t > 0) \xrightarrow[x \to \infty]{} T_0; \tag{8.55}$$

$$T(x, t = 0) = T_0. \tag{8.56}$$

The analytical solution of which is expressed quite simply as:

$$\frac{T - T_w}{T_0 - T_w} = \text{erf}\left(\frac{\psi}{2}\right),$$ (8.57)

where $\psi = x/\sqrt{\mathcal{D}t}$ and erf is the error function. If subject to a surface convective boundary condition (with heat-transfer coefficient α), the solution would be:

$$\frac{T - T_b}{T_0 - T_b} = \mathcal{E}(\psi, \beta), \quad \text{with} \quad \psi = \frac{x}{\sqrt{\mathcal{D}t}} \quad \text{and} \quad \beta = \frac{\alpha\sqrt{\mathcal{D}t}}{\lambda},$$ (8.58)

where the function $\mathcal{E}$ is expressed as (with $\text{erfc}(x) = 1 - \text{erf}(x)$):

$$\mathcal{E}(\psi, \beta) = \text{erf}\left(\frac{\psi}{2}\right) + \exp\left(\psi\beta + \beta^2\right)\text{erfc}\left(\frac{\psi}{2} + \beta\right).$$ (8.59)

8.5.3   Dynamic Modelling of Latent Heat Storage

Heat storage within liquid–solid PCMs involves not only diffusion (and potentially advection) within liquid and solid phases, but also the propagation of melting fronts. Reliable estimates of the LHS storage dynamics, even though complicated than that in SHS systems, can be provided by simplified modelling tools in lieu of complex and time-consuming three-dimensional CFD simulations.

Like in SHS systems, the first law of thermodynamics applied to a PCM body of mass m at a temperature $T_{st}(\mathbf{x}, t)$, heated or cooled by an external heat-transfer fluid flow at T_b, provides us with an equation to describe the time-resolved evolution of the system:

$$m\frac{d\langle h_{st}\rangle}{dt} = \alpha S(T_b - \langle T_{st}\rangle),$$ (8.60)

where $\langle h_{st}\rangle$ denotes the mass-averaged specific enthalpy of the PCM:

$$\langle h_{st}\rangle(t) = \frac{1}{m}\int_V \rho h_{st}(\mathbf{x}, t)\,d\mathbf{x}.$$ (8.61)

As such, Equation (8.60) does not provide useful insight into the transient heat transfer. Let us then consider the so-called two-phase Lamé–Clapeyron–Stefan problem (referred to as the Stefan problem), which describes the one-dimensional propagation of a a two-phase interface within a PCM, to understand the melting/solidification dynamics.

One-Dimensional Stefan Problem

The propagation rate of the melting front, $\dot{x}_{mf}(t)$, in a semi-infinite slab, initially fully solidified at a uniform temperature, T_0, and subject to a sudden change in surface temperature, obeys a set of equations that describe heat diffusion:

In the solid phase

$$\frac{\partial T_s}{\partial t} = \mathcal{D}_s \frac{\partial^2 T_s}{\partial x^2} \qquad\qquad x > x_{mf}(t), \ t > 0 \qquad (8.62)$$

B.C. $T_s(x = x_{mf}, t) = T_m$ $\qquad\qquad\qquad t > 0$ $\qquad (8.63)$

B.C. $T_s(x, t) \xrightarrow[x \to \infty]{} T_0$ $\qquad\qquad x > x_{mf}(t), \ t > 0$ $\qquad (8.64)$

I.C. $T_s(x, t = 0) = T_0$ $\qquad\qquad x > x_{mf}(t = 0).$ $\qquad (8.65)$

In the liquid phase

$$\frac{\partial T_l}{\partial t} = \mathcal{D}_l \frac{\partial^2 T_l}{\partial x^2} \qquad\qquad x < x_{mf}(t), \ t > 0 \qquad (8.66)$$

B.C. $T_l(x = 0, t) = T_w$ $\qquad\qquad x = 0, \ t > 0$ $\qquad (8.67)$

B.C. $T_l(x = x_{mf}, t) = T_m$ $\qquad\qquad\qquad t > 0.$ $\qquad (8.68)$

And the enthalpy balance at the solid–liquid interface provides us with an additional equation to predict the moving interface velocity:

$$\lambda_s \left(\frac{\partial T_s}{\partial x}\right)_{x=x_{mf}} - \lambda_l \left(\frac{\partial T_l}{\partial x}\right)_{x=x_{mf}} = \rho \Delta h_{ls} \dot{x}_{mf}. \qquad (8.69)$$

A closed-form solution to the problem defined by Equations (8.62) to (8.68) exists and is easily derived. Let us first note that the liquid- and solid-phase problems are 'moving-frame' semi-infinite heat-diffusion problems with a sudden change in surface temperature – the 'static' solution of which was given in Equation (8.57). By imposing a solution of the same form, the so-called Neumann solution is obtained:

$$\frac{T_s - T_0}{T_m - T_0} = \frac{\text{erfc}\,(\psi_s)}{\text{erfc}\,(\psi_{s,mf})}, \qquad \psi_s(x, t) = \frac{x}{2\sqrt{\mathcal{D}_s t}} \qquad (8.70)$$

$$\frac{T_l - T_w}{T_m - T_w} = \frac{\text{erf}\,(\psi_l)}{\text{erf}\,(\psi_{l,mf})}, \qquad \psi_l(x, t) = \frac{x}{2\sqrt{\mathcal{D}_l t}}, \qquad (8.71)$$

where $\psi_{s/l,mf}(t) = \psi_{s/l}(x_{mf}(t), t)$. A transcendental equation is obtained to predict the melting interface position, $x_{mf}(t)$, by plugging the above solutions for the liquid and solid temperatures into Equation (8.69):

$$\sqrt{\pi}\,\psi_{s,mf} = \kappa \text{Ste}_l \frac{e^{-\psi_{l,mf}^2}}{\text{erf}(\psi_{l,mf})} - \text{Ste}_s \frac{e^{-\psi_{s,mf}^2}}{\text{erfc}(\psi_{s,mf})}, \quad \text{with} \qquad (8.72)$$

$$\kappa = \sqrt{\frac{\mathcal{D}_l}{\mathcal{D}_s}}, \quad \text{Ste}_s = \frac{c_{p,s}\,(T_m - T_0)}{\Delta h_{ls}}, \quad \text{and} \quad \text{Ste}_l = \frac{c_{p,l}\,(T_w - T_m)}{\Delta h_{ls}}. \qquad (8.73)$$

Following the analysis performed by Zubair and Chaudhry (1994) and Tarzia (1982), an analytical solution to the one-dimensional Stefan problem with a *convective boundary condition* can be derived for specific time-dependent heat-transfer coefficients (which must be proportional to the inverse of the square root of time). This solution is of little interest for direct modelling purposes due to the said constraint on the form of the boundary condition; the formulas are thus not explicitly given here.

Yet, it is worth mentioning this particular result, which can provide an exact solution to be used as a benchmark to verify the accuracy of various numerical methods.

Numerical Methods for Phase-Change Problems

The advanced numerical approaches required for thermodynamically consistent predictions of multidimensional transient phase-change processes can be classified into two main categories: (i) fixed-domain formulation; and (ii) variable-domain formulation (Jana, Ray & Durst 2007). In the latter, advection-diffusion equations are solved in each of the phases, while enthalpy and mass balances are written at the interface (solid–liquid for melting/solidification problems) to determine the phase-change front velocity. In fixed-grid formulations, on the other hand, the latent heat evolution is accounted for in the single set of governing equations that are used to solve both phases.

A detailed presentation of moving-boundary techniques for phase-change problems is proposed by Crank (1987). These numerical methods are not described further in this textbook, the purpose of which is simply to present reduced-order modelling tools.

8.6   Summary

In the context of heat utilisation, thermal energy storage (TES) systems provide a customisable and configurable solution to mitigate the detrimental effects arising from heat-source intermittency. In order to effectively utilise time-varying heat sources that exhibit irregular alternations (such as the solar heat, which can suddenly drop due to a passing cloud, and which obviously does not provide any heat input during night time), heat-storage media act as thermal buffers that can overcome the demand-to-supply mismatch and mitigate the performance degradation of heat-utilisation technologies due to transient effects. In this chapter, simple tools have been presented to assess the potential of various kinds of TES techniques, measured through a set of key techno-economic performance indicators, defined in Section 8.1.1.

Reduced-order modelling tools have been presented to estimate the storage (energy) and power capacity – amongst other indicators – of the three main categories of TES media:

- sensible heat-storage media, both liquid and solid, which offer relatively low-storage density but remain attractive due to easy-to-solve storage constraints and low investment costs;
- latent heat-storage media, especially liquid–solid PCMs that offer moderate to high storage density, which, together with the relatively low thermal conductivity of PCMs, make the development of cost-effective compact thermal batteries challenging;
- (thermo-)Chemical materials, in which energy is stored and released through breaking and reforming molecular bonds in reversible endo-/exothermic reactions, characterised by (ultra-)highenergy storage densities.

The characteristics and potential applications of, and practical aspects related to, sensible, latent and chemical TES have been discussed in detail in Sections 8.2, 8.3 and 8.4, respectively. Finally, a series of analytical and numerical techniques have been presented in section 8.5, along with adequate heat-transfer correlations, to predict the performance and dynamic characteristics of heat-storage solutions using several containment techniques (encapsulation, packed bed of pebbles, coils, etc.).

Appendix: Commercial Systems

Table A.1 Websites for the companies offering different types of waste-heat recovery technology. Websites last accessed on 9 February 2017

Technology	Manufacturer	Website
Steam	Elliott	www.elliott-turbo.com/
	Heliex Power	www.heliexpower.com/
	MAN	
	Siemens	www.energy.siemens.com/
	Spilling	www.spilling.info/
	VOITH	www.voith.com/en/
ORC	Atlas Copco	www.atlascopco-gap.com/
	BEP	www.e-rational.net/
	Calnetix	www.calnetix.com/
	DeVeTec	www.devetec.com/
	Electratherm	https://electratherm.com/
	Enertime	www.enertime.com/en/
	Enogia	www.enogia.com/
	Exergy	http://exergy-orc.com/
	GE	https://powergen.gepower.com/
	Maxxtec	www.maxxtec.com/
	ORMAT	www.ormat.com/
	Siemens	www.energy.siemens.com/
	Triogen	www.triogen.nl/
	Turboden	www.turboden.eu/en/
Heat pump	Daikin	http://lit.daikinapplied.com/
	Emerson	www.emersonclimate.com/
	Mayekawa	www.mayekawa.com/
	Roth	www.roth-usa.com/
	Versatec	www.waterfurnace.com/
Absorption chiller	Ago	www.ago.ag/en/
	Carrier	www.carrier.com/
	Cention	http://centioncorp.com/
	LG	www.lg.com/
	Thermax	www.thermax-europe.com/
	Trane	www.trane.com/
	York	www.johnsoncontrols.com/

References

Abhat, A. (1983), 'Low temperature latent heat thermal energy storage: heat storage materials', *Emerson Climate Technologies, Delta-ee 3rd Annual Heat Pumps & Utilities Roundtable* **30**(4), 313–332.

Abrahamsson, K. & Jernqvist, Å. (1993), 'Carnot comparison of multi-temperature level absorption heat cycles', *International Journal of Refrigeration* **16**(4), 240–246.

Alshammari, F., Pesyridis, A., Karvountzis-Kontakiotis, A., Franchetti, B. & Pesmazoglou, Y. (2018), 'Experimental study of a small-scale organic Rankine cycle waste heat recovery system for a heavy duty diesel engine with focus on the radial inflow turbine expander performance', *Applied Energy* **215**, 543–555.

Alshammari, F., Usman, M. & Pesyridis, A. (2018), 'Expanders for organic Rankine cycle technology', in E. Wang, ed., *Organic Rankine cycle technology for heat recovery*, InTech Open, pp. 41–59.

Andreasen, J., Kærn, M., Pierobon, L., Larsen, U. & Haglind, F. (2016), 'Multi-objective optimization of organic Rankine cycle power plants using pure and mixed working fluids', *Energies* **9**(5), 322.

Angelino, G. (1968), 'Carbon dioxide condensation cycles for power production', *Journal of Engineering for Gas Turbines and Power* **90**(3), 287–295.

Angelino, G. & Colonna, P. (1998), 'Multicomponent working fluids for organic Rankine cycles (ORCs)', *Energy* **23**(6), 449–463.

Azzolin, M., Bortolin, S. & Del Col, D. (2016), 'Flow boiling heat transfer of a zeotropic binary mixture of new refrigerants inside a single microchannel', *International Journal of Thermal Sciences* **110**, 83–95.

Backhaus, S. & Swift, G. W. (2002), 'A thermoacoustic-Stirling heat engine: detailed study', *The Journal of the Acoustical Society of America* **107**(6), 3148–3166.

Badr, O., O'Callaghan, P., Hussein, M. & Probert, S. (1984), 'Multi-vane expanders as prime movers for low-grade energy organic Rankine-cycle engines', *Applied Energy* **16**(1), 129–146.

Badr, O., Probert, S. D. & O'Callaghan, P. W. (1985), 'Selecting a working fluid for a Rankine-cycle engine', *Applied Energy* **21**(1), 1–42.

Basecq, V., Michaux, G., Inard, C. & Blondeau, P. (2013), 'Short-term storage systems of thermal energy for buildings: a review', *Advances in Building Energy Research* **7**(1), 66–119.

BCS Incorporated. (2008), 'Waste heat recovery: technology and opportunities in U.S. industry'. Prepared for the U.S. Department of Energy, Industrial Technologies Program. www1.eere.energy.gov/manufacturing/intensiveprocesses/pdfs/waste_heat_recovery.pdf.

Bell, I. H., Wronski, J., Quoilin, S. & Lemort, V. (2014), 'Pure and pseudo-pure fluid thermo-physical property evaluation and the open-source thermophysical property library coolprop', *Industrial and Engineering Chemistry Research* **53**(6), 2498–2508.

Berntsson, T. & Åsblad, A. (2015), 'Annex XV: industrial excess heat recovery–technologies and applications'. Prepared for the International Energy Agency, Industrial Energy-Related Technologies and Systems (IETS) Technology Collaboration Programme. https://iea-industry.org/app/uploads/Final-report-IETS-Annex-15-Task-2.pdf.

Bianchi, G., Panayiotou, G. P., Aresti, L. et al. (2019), 'Estimating the waste heat recovery in the European Union industry', *Energy, Ecology and Environment* **4**(5), 211–221.

Bisio, G. & Rubatto, G. (2000), 'Energy saving and some environment improvements in coke-oven plants', *Energy* **25**(3), 247–265.

Blanchard, C. H. (1980), 'Coefficient of performance for finite speed heat pump', *Journal of Applied Physics* **51**(5), 2471–2472.

Brignoli, R. & Brown, J. S. (2015), 'Organic Rankine cycle model for well-described and not-so-well-described working fluids', *Energy* **86**, 93–104.

Brückner, S., Liu, S., Miró, L. et al. (2015), 'Industrial waste heat recovery technologies: an economic analysis of heat transformation technologies', *Applied Energy* **151**, 157–167.

Brueckner, S., Miró, L., Cabeza, L. F., Pehnt, M. & Laevemann, E. (2014), 'Methods to estimate the industrial waste heat potential of regions – a categorization and literature review', *Renewable and Sustainable Energy Reviews* **38**, 164–171.

Brun, K., Friedman, P. & Dennis, R. (2017), *Fundamentals and applications of supercritical carbon dioxide (sCO2) based power cycles*, Woodhead Publishing.

Campana, F., Bianchi, M., Branchini, L. et al. (2013), 'ORC waste heat recovery in European energy intensive industries: energy and GHG savings', *Energy Conversion and Management* **76**(1), 244–252.

Campana, F., Bianchi, M., Branchini, L. et al. (2013*b*), 'ORC waste heat recovery in European energy intensive industries: energy and GHG savings', *Energy Conversion and Management* **76**, 244–252.

Chan, C. W., Ling-Chin, J. & Roskilly, A. P. (2013), 'Reprint of "a review of chemical heat pumps, thermodynamic cycles and thermal energy storage technologies for low grade heat utilisation", *Applied Thermal Engineering* **53**(2), 160–176.

Chapman, W. G., Gubbins, K. E., Jackson, G. & Radosd, M. (1990), 'New reference equation of state for associating liquids', *Industrial & Engineering Chemistry Research* **29**(8), 1709–1721.

Chen, H., Goswami, D. Y. & Stefanakos, E. K. (2010), 'A review of thermodynamic cycles and working fluids for the conversion of low-grade heat', *Renewable and Sustainable Energy Reviews* **14**(1), 3059–3067.

Chen, J. (1995), 'The equivalent cycle system of an endoreversible absorption refrigerator and its general performance characteristics', *Energy* **20**(10), 995–1003.

Chintala, V., Kumar, S. & Pandey, J. K. (2018), 'A technical review on waste heat recovery from compression ignition engines using organic Rankine cycle', *Renewable and Sustainable Energy Reviews* **81**, 493–509.

Chua, K. J., Chou, S. K. & Yang, W. M. (2010), 'Advances in heat pump systems: a review', *Applied Energy* **87**(12), 3611–3624.

Cignitti, S., Andreasen, J. G., Haglind, F., Woodley, J. M. & Abildskov, J. (2017), 'Integrated working fluid-thermodynamic cycle design of organic Rankine cycle power systems for waste heat recovery', *Applied Energy* **203**, 442–453.

Colonna, P., Casati, E., Trapp, C. et al. (2015), 'Organic Rankine cycle power systems: from the concept to current technology, applications and an outlook to the future', *Journal of Engineering for Gas Turbines and Power* **137**(10), 19.

Costall, A. W., Hernandez, A. G., Newton, P. J. & Martinez-Botas, R. F. (2015), 'Design methodology for radial turbo expanders in mobile organic Rankine cycle applications', *Applied Energy* **157**, 729–743.

Crank, J. (1987), *Free and moving boundary problems*, Oxford University Press.

Dai, Y., Wang, J. & Gao, L. (2009), 'Parametric optimization and comparative study of organic Rankine cycle (ORC) for low grade waste heat recovery', *Energy Conversion and Management* **50**(1), 576–582.

de Beer, J., Worrell, E. & Blok, K. (1998), 'Future technologies for energy-efficient iron and steel making', *Annual Review of Energy and the Environment* **23**(1), 123–205.

Dittus, F. & Boelter, L. (1985), 'Heat transfer in automobile radiators of the tubler type', *Univ. Calif. Pubs. Eng.* **12**(1), 3–22

Dixon, S. L. (2005), *Fluid mechanics and thermodynamics of turbomachinery*, 5th ed., Butterworth-Heinemann.

Dostal, V. (2004), '*A supercritical carbon dioxide cycle for next generation nuclear reactors*', PhD, Massachusetts Institute of Technology. http://dspace.mit.edu/handle/1721.1/17746

Eicker, U. & Pietruschka, D. (2009), 'Design and performance of solar powered absorption cooling systems in office buildings', *Energy and Buildings* **41**(1), 81–91.

Element Energy Limited. (2014), 'The potential for recovering and using surplus heat from industry'. Led by Element Energy Limited, prepared by Ecofys, Imperial College London, Paul Stevenson and Robert Hyde for the Department of Energy and Climate Change, London. https://assets.publishing.service.gov.uk/government/uploads/system/uploads/attachment_data/file/294900/element_energy_et_al_potential_for_recovering_and_using_surplus_heat_from_industry.pdf.

Elsheniti, M. B., Elsamni, O. A., Al-dadah, R. K. et al. (2017), 'Adsorption refrigeration technologies', in C. Ghenai & T. Salameh, eds, *Sustainable air conditioning systems*, IntechOpen, chapter 10.

Elson, A., Hampson, A. & Tidball, R. (2015), *Waste heat to power market assessment*, number March.

Energetics Incorporated. (2000), 'Energy and environmental profile of the U.S. iron and steel industry'. Prepared for the U.S. Department of Energy, Office of Industrial Technologies. www.energy.gov/sites/prod/files/2013/11/f4/steel_profile.pdf.

Entec UK Limited. (2006), 'Cost estimation methodology: the marine energy challenge approach of energy produced by marine energy systems'. Commissioned by the Carbon Trust. www.carbontrust.com/resources/marine-cost-of-energy-methodology.

Esposito, M. C., Pompini, N., Gambarotta, A. et al. (2015), 'Nonlinear model predictive control of an organic Rankine cycle for exhaust waste heat recovery in automotive engines', *IFAC-PapersOnLine* **48**(15), 411–418.

European Commission. (2008), 'Guide to cost benefit analysis of investment projects'. http://ec.europa.eu/regional_policy/sources/docgener/guides/cost/guide2008_en.pdf.

Evans, O. (1805), *The abortion of the young steam engineer's guide*, The Oliver Evans Press.

Fallahi, A., Guldentops, G., Tao, M., Granados-Focil, S. & Van Dessel, S. (2017), 'Review on solid-solid phase change materials for thermal energy storage: molecular structure and thermal properties', *Applied Thermal Engineering* **127**, 1427–1441.

Feher, E. G. (1968), 'The supercritical thermodynamic power cycle', *Energy Conversion* **8**(2), 85–90.

Fernández, A., Galleguillos, H., Fuentealba, E. & Pérez, F. (2015), 'Thermal characterization of hitec molten salt for energy storage in solar linear concentrated technology', *Journal of Thermal Analysis and Calorimetry* **122**(1), 3–9.

Fischer, J. (2011), 'Comparison of trilateral cycles and organic Rankine cycles', *Energy* **36**(10), 6208–6219.

Ge, H., Li, H., Mei, S. & Liu, J. (2013), 'Low melting point liquid metal as a new class of phase change material: an emerging frontier in energy area', *Renewable and Sustainable Energy Reviews* **21**, 331–346.

Gnielinski, V. (1975), 'Neue gleichungen für den wärme-und den stoffübergang in turbulent durchströmten rohren und kanälen', *Forschung im Ingenieurwesen A* **41**(1), 8–16.

Hadorn, J.-C. (2008), 'Advanced storage concepts for active solar energy – IEA SHC task 32 2003–2007', in *Eurosun 1st International Conference on Solar Heating, Cooling and Buildings*, 7–10 October, Lisbon pp. 1–8.

Hepbasli, A. & Kalinci, Y. (2009), 'A review of heat pump water heating systems', *Renewable and Sustainable Energy Reviews* **13**(6–7), 1211–1229.

Herold, K. E., Radermacher, R. & Klein, S. A. (2016), *Absorption chillers and heat pumps*, ESDU.

Hewitt, G. (1994), 'Selection and costing of heat exchangers', Technical report, ESDU 92013.

Höhlein, S., König-Haagen, A. & Brüggemann, D. (2018), 'Macro-encapsulation of inorganic phase-change materials (pcm) in metal capsules', *Materials* **11**(9), 1752.

Horuz, I. (1998), 'A comparison between ammonia-water and water-lithium bromide solutions in vapor absorption refrigeration systems', *International Communications in Heat and Mass Transfer* **25**(5), 711–721.

Hui, L., N'Tsoukpoe, K. E., Lingai, L. et al. (2011), 'Evaluation of a seasonal storage system of solar energy for house heating using different absorption couples', *Energy Conversion and Management* **52**(6), 2427–2436.

Internation Energy Agency. (2011), 'Solar energy perspectives'. https://web.archive.org/web/20120113032718/http://www.iea.org/Textbase/npsum/solar2011SUM.pdf.

International Renewable Energy Agency. (2013), 'Thermal energy storage: technology brief E17'. www.irena.org/-/media/Files/IRENA/Agency/Publication/2013/IRENA-ETSAP-Tech-Brief-E17-Thermal-Energy-Storage.pdf.

Jakobs, R., Cibis, D. & Laue, H. J. (2010), 'Status and outlook: industrial heat pumps', in *International Refrigeration and Air Conditioning Conference*, pp. 1–8.

Jana, S., Ray, S. & Durst, F. (2007), 'A numerical method to compute solidification and melting processes', *Applied Mathematical Modelling* **31**(1), 93–119.

Johnson, I. & Choate, W. T. (2008) , 'Waste heat recovery: technology and opportunities in U.S. industry', Technical report, BCS Incorporated.

Kalina, A. (1984), 'Combined-cycle system with novel bottoming cycle', *Journal of Engineering for Gas Turbines and Power* **106**(4), 737–742.

Karellas, S., Schuster, A. & Leontaritis, A.-D. (2012), 'Influence of supercritical ORC parameters on plate heat exchanger design', *Applied Thermal Engineering* **33–34**(1), 70–76.

Kirmse, C. J., Oyewunmi, O. A., Taleb, A. I., Haslam, A. J. & Markides, C. N. (2017), 'A two-phase single-reciprocating-piston heat conversion engine: non-linear dynamic modelling', *Applied Energy* **186**, 359–375.

Kongtragool, B. & Wongwises, S. (2003), 'A review of solar-powered Stirling engines and low temperature differential Stirling engines', *Renewable and Sustainable Energy Reviews* **7**(2), 131–154.

Kraus, A. D. (2003), 'Heat exchangers', in A. Bejan & A. D. Kraus, eds., *Heat transfer handbook*, John Wiley, pp. 797–911.

Lampe, M., Stavrou, M., Bu, H. M., Gross, J. & Bardow, A. (2014), 'Simultaneous optimization of working fluid and process for organic Rankine cycles using PC-SAFT', *Industrial & Engineering Chemistry Research* **53**, 8821–8830.

Lampe, M., Stavrou, M., Schilling, J. et al. (2015), 'Computer-aided molecular design in the continuous-molecular targeting framework using group-contribution PC-SAFT', *Computers and Chemical Engineering* **81**, 278–287.

Larsen, U., Pierobon, L., Haglind, F. & Gabrielii, C. (2013), 'Design and optimisation of organic Rankine cycles for waste heat recovery in marine applications using the principles of natural selection', *Energy* **55**, 803–812.

Le, V. L., Feidt, M., Kheiri, A. & Pelloux-prayer, S. (2014), 'Performance optimization of low-temperature power generation by supercritical ORCs (organic Rankine cycles) using low GWP (global warming potential) working fluids', *Energy* **67**, 513–526.

Lecompte, S., Ameel, B., Ziviani, D., Van Den Broek, M. & De Paepe, M. (2014), 'Exergy analysis of zeotropic mixtures as working fluids in organic Rankine cycles', *Energ Convers Manage* **85**, 727–739.

Lecompte, S., Huisseune, H., van den Broek, M., De Schampheleire, S. & De Paepe, M. (2013), 'Part load based thermo-economic optimization of the organic Rankine cycle (ORC) applied to a combined heat and power (CHP) system', *Applied Energy* **111**, 871–881.

Lecompte, S., Huisseune, H., van den Broek, M., Vanslambrouck, B. & De Paepe, M. (2015), 'Review of organic Rankine cycle (ORC) architectures for waste heat recovery', *Renewable and Sustainable Energy Reviews* **47**, 448–461.

Lemmens, S. (2016), 'Cost engineering techniques & their applicability for cost estimation of organic Rankine cycle systems', *Energies* **9**(7), 485.

Lemmon, E., Huber, M. & McLinden, M. (2013), 'NIST standard reference database 23: reference fluid thermodynamic and transport properties-REFPROP', Version 10.0, National Institute of Standards and Technology, Standard Reference Data Program, Gaithersburg, 2018.

Lemort, V. & Legros, A. (2017), 'Positive displacement expanders for organic Rankine cycle systems', in E. Macchi & M. Astolfi, eds., *Organic Rankine cycle (ORC) power systems*, Elsevier, pp. 361–396.

Lienhard, J. H. (2019), *A heat transfer textbook*, Dover.

Löf, G. & Hawley, R. (1948), 'Unsteady-state heat transfer between air and loose solids', *Industrial & Engineering Chemistry* **40**(6), 1061–1070.

Ma, G. Y., Cai, J. J., Zeng, W. W. & Dong, H. (2012), 'Analytical research on waste heat recovery and utilization of China's iron & steel industry', *Energy Procedia* **14**, 1022–1028.

Maizza, V. & Maizza, A. (1996), 'Working fluids in non-steady flows for waste energy recovery systems', *Applied Thermal Engineering* **16**(7), 579–590.

Markides, C. N. (2015), 'Low-concentration solar-power systems based on organic Rankine cycles for distributed-scale applications: overview and further developments', *Frontiers in Energy Research* **3**(December), 1–16.

McKenna, R. C. (2009), *'Industrial energy efficiency: Interdisciplinary perspectives on the thermodynamic, technical and economic constraints'*, Phd, University of Bath.

Nellissen, P. & Wolf, S. (2015), 'Heat pumps in non-domestic applications in Europe: potential for an energy revolution', Delta-ee 3rd Annual Heat Pumps & Utilities Roundtable, *Solar Energy*.

Novikov, I. I. (1958), 'The efficiency of atomic power stations (a review)', *Journal of Nuclear Energy II* **7**(1–2), 125–128.

Oluleye, G., Jobson, M. & Smith, R. (2015), 'A hierarchical approach for evaluating and selecting waste heat utilization opportunities', *Energy* **90**, 5–23.

Oluleye, G., Jobson, M., Smith, R. & Perry, S. J. (2016), 'Evaluating the potential of process sites for waste heat recovery', *Applied Energy* **161**, 627–646.

Ommen, T., Jensen, J. K., Markussen, W. B., Reinholdt, L. & Elmegaard, B. (2015), 'Technical and economic working domains of industrial heat pumps: part 1 – single stage vapour compression heat pumps', *International Journal of Refrigeration* **55**, 168–182.

Orr, B., Akbarzadeh, A., Mochizuki, M. & Singh, R. (2016), 'A review of car waste heat recovery systems utilising thermoelectric generators and heat pipes', *Applied Thermal Engineering* **101**, 490–495.

Oyewunmi, O. A., Lecompte, S., De Paepe, M. & Markides, C. N. (2017), 'Thermoeconomic analysis of recuperative sub- and transcritical organic Rankine cycle systems', *Energy Procedia* **129**, 58–65.

Oyewunmi, O. A. & Markides, C. N. (2016), 'Thermo-economic and heat transfer optimization of working-fluid mixtures in a low-temperature organic Rankine cycle system', *Energies* **9**(6), 448.

Oyewunmi, O. A., Taleb, A. I., Haslam, A. J. & Markides, C. N. (2016), 'On the use of SAFT-VR Mie for assessing large-glide fluorocarbon working-fluid mixtures in organic Rankine cycles', *Applied Energy* **163**, 263–282.

Papadopoulos, A. I., Stijepovic, M. & Linke, P. (2010), 'On the systematic design and selection of optimal working fluids for organic Rankine cycles', *Applied Thermal Engineering* **30**(6–7), 760–769.

Papadopoulos, A. I., Stijepovic, M., Linke, P., Seferlis, P. & Voutetakis, S. (2013), 'Toward optimum working fluid mixtures for organic Rankine cycles using molecular design and sensitivity analysis', *Industrial & Engineering Chemistry Research* **52**, 12116–12133.

Papapetrou, M., Kosmadakis, G., Cipollina, A., La Commare, U. & Micale, G. (2018), 'Industrial waste heat: estimation of the technically available resource in the EU per industrial sector, temperature level and country', *Applied Thermal Engineering* **138**, 207–216.

Pehnt, M., Bödeker, J., Arens, M., Jochem, E. & Idrissova, F. (2010), 'Die Nutzung industrieller Abwärme – technisch-wirtschaftliche Potenziale und energiepolitische Umsetzung'. Bericht im Rahmen des Vorhabens Wissenschaftliche Begleitforschung zu übergreifenden technischen, ökologischen, ökonomischen und strategischen Aspekten des nationalen Teils der Klimaschutzinitiative. Institut für Energie-und Umweltforschung (ifeu), Heidelberg. www.ifeu.de/wp-content/uploads/Nutzung_industrieller_Abwaerme.pdf.

Peng, D.-Y. & Robinson, D. B. (1976), 'A new two-constant equation of state', *Industrial & Engineering Chemistry Fundamentals* **15**(1), 59–64.

Perkins, J. (1835), 'Apparatus and means for producing ice and in cooling fluids'.

Persico, G. & Pini, M. (2017), 'Fluid dynamic design of organic Rankine cycle turbines', in E. Macchi & M. Astolfi, eds., *Organic Rankine cycle (ORC) power systems*, Elsevier, pp. 253–297.

Petukhov, B. (1970), 'Heat transfer and friction in turbulent pipe flow with variable physical properties', in *Advances in heat transfer*, Vol. 6, Elsevier, pp. 503–564.

Pierobon, L., Nguyen, T. V., Larsen, U., Haglind, F. & Elmegaard, B. (2013), 'Multi-objective optimization of organic Rankine cycles for waste heat recovery: application in an offshore platform', *Energy* **58**(1), 538–549.

Pinel, P., Cruickshank, C. A., Beausoleil-Morrison, I. & Wills, A. (2011), 'A review of available methods for seasonal storage of solar thermal energy in residential applications', *Renewable and Sustainable Energy Reviews* **15**(7), 3341–3359.

Poling, B. E., Prausnitz, J. M. & O'Connell, J. P. (2001), *The properties of gases and liquids*, 5th ed., McGraw-Hill.

Quintard, M. & Whitaker, S. (1994), 'Transport in ordered and disordered porous media II: generalized volume averaging', *Transport in porous media* **14**(2), 179–206.

Quoilin, S., Declaye, S., Tchanche, B. F. & Lemort, V. (2011), 'Thermo-economic optimization of waste heat recovery organic Rankine cycles', *Applied Thermal Engineering* **31**(14–15), 2885–2893.

Quoilin, S., van den Broek, M., Declaye, S., Dewallef, P. & Lemort, V. (2013), 'Techno-economic survey of organic Rankine cycle (ORC) systems', *Renewable and Sustainable Energy Reviews* **22**(1), 168–186.

Rahbar, K., Mahmoud, S., Al-Dadah, R. K., Moazami, N. & Mirhadizadeh, S. A. (2017), 'Review of organic Rankine cycle for small-scale applications', *Energy Conversion and Management* **134**, 135–155.

Sarbu, I. & Sebarchievici, C. (2018), 'A comprehensive review of thermal energy storage', *Sustainability* **10**(1), 191.

Schilling, J., Lampe, M. & Bardow, A. (2017), '1-stage CoMT-CAMD: an approach for integrated design of ORC process and working fluid using PC-SAFT', *Chemical Engineering Science*, 159, 217–30.

Schilling, J., Tillmanns, D., Lampe, M. et al. (2017), 'From molecules to dollars: integrating molecular design into thermo-economic process design using consistent thermodynamic modeling', *Molecular Systems Design and Engineering* **2**(3), 301–320.

Schuster, A., Karellas, S. & Aumann, R. (2010), 'Efficiency optimization potential in supercritical organic Rankine cycles', *Energy* **35**(2), 1033–1039.

Seider, W. D., Seader, J. D. & Lewin, D. R. (2009), *Product & process design principles: synthesis, analysis and evaluation*, John Wiley.

Sharma, A., Tyagi, V. V., Chen, C. & Buddhi, D. (2009), 'Review on thermal energy storage with phase change materials and applications', *Renewable and Sustainable Energy Reviews* **13**(2), 318–345.

Shu, G., Liang, Y., Wei, H. et al. (2013), 'A review of waste heat recovery on two-stroke IC engine aboard ships', *Renewable and Sustainable Energy Reviews* **19**, 385–401.

Sieder, E. & Tate, G. (1936), 'Heat transfer and pressure drop of liquids in tubes', *Industrial & Engineering Chemistry* **28**(12), 1429–1435.

Smith, I. K. (1993), 'Development of the trilateral flash cycle system: part 1: fundamental considerations', *Proceedings of the Institution of Mechanical Engineers, Part A: Journal of Power and Energy* **207**(3), 179–194.

Smith, I. K. & da Silva, R. P. M. (1994), 'Development of the trilateral flash cycle system: part 2: increasing power output with working fluid mixtures', *Proceedings of the Institution of Mechanical Engineers, Part A: Journal of Power and Energy* **208**(2), 135–144.

Smith, I. K., Stosic, N. & Aldis, C. A. (1996), 'Development of the trilateral flash cycle system: part 3: the design of high-efficiency two-phase screw expanders', *Proceedings of the Institution of Mechanical Engineers, Part A: Journal of Power and Energy* **210**(1), 75–93.

Smith, I. K., Stosic, N. & Kovacevic, A. (2005), 'Screw expanders increase output and decrease the cost of geothermal binary power plant systems', in *Geothermal Resources Council Annual Meeting*, Reno.

Smith, I. K., Stosic, N. & Kovacevic, A. (2014), *Power recovery from low grade heat by means of screw expanders*, Chandos.

Smith, R. (2005), *Chemical process design and integration*, 2nd ed., John Wiley.

Smith, T. C. B. (2012), 'Power dense thermofluidic oscillators for high load applications', in *2nd International Energy Conversion Engineering Conference*, 16–19 August, Providence.

Soave, G. (1972), 'Equilibrium constants from a modified Redlich-Kwong equation of state', *Chemical Engineering Science* **27**(6), 1107–1203.

Song, J., Song, Y. & Gu, C.-w. (2015), 'Thermodynamic analysis and performance optimization of an organic Rankine cycle (orc) waste heat recovery system for marine diesel engines', *Energy* **82**, 976–985.

Sprouse, C. & Depcik, C. (2013), 'Review of organic Rankine cycles for internal combustion engine exhaust waste heat recovery', *Applied Thermal Engineering* **51**(1–2), 711–722.

Srikhirin, P., Aphornratana, S. & Chungpaibulpatana, S. (2001), 'A review of absorption refrigeration technologies', *Renewable and Sustainable Energy Reviews* **5**, 343–372.

Stammers, C. W. (1979), 'The operation of the Fluidyne heat engine at low differential temperatures', *Journal of Sound and Vibration* **63**(4), 507–516.

Stouffs, P., Tazerout, M. & Wauters, P. (2001), 'Thermodynamic analysis of reciprocating compressors', *International Journal of Thermal Sciences* **40**(1), 52–66.

Su, W., Zhao, L. & Deng, S. (2017), 'Simultaneous working fluids design and cycle optimization for organic Rankine cycle using group contribution model', *Applied Energy* **202**, 618–627.

Tarzia, D. A. (1982), 'An inequality for the coefficient σ of the free boundary $s(t) = 2\sigma\sqrt{t}$ of the Neumann solution for the two-phase Stefan problem', *Quarterly of Applied Mathematics* **39**(4), 491–497.

Tauveron, N., Colasson, S. & Gruss, J. (2014), 'Available systems for the conversion of waste heat to electricity', in *Proceedings of the ASME 2014 International Mechanical Engineering Congress & Exposition*, 14–20 November, Montreal,

Tauveron, N., Colasson, S. & Gruss, J. (2015), 'Conversion of waste heat to electricity: cartography of possible cycles due to hot source characteristics', in *Proceedings of ECOS 2015 – The 28th International Conference on Efficiency, Cost, Optimization, Simulation and Environmental Impact of Energy Systems*, 30 June–3 July, Pau, pp. 1–13.

Tchanche, B. F., Lambrinos, G., Frangoudakis, A. & Papadakis, G. (2011), 'Low-grade heat conversion into power using organic Rankine cycles – a review of various applications', *Renewable and Sustainable Energy Reviews* **15**(8), 3963–3979.

Tijani, M. E. & Spoelstra, S. (2011), 'A high performance thermoacoustic engine', *Journal of Applied Physics* **110**(9), 09351.

Tona, P. & Peralez, J. (2015), 'Control of organic Rankine cycle systems on board heavy-duty vehicles: a survey', *IFAC-PapersOnLine* **48**(15), 419–426.

Turton, R., Bailie, R., Whiting, W. & Shaelwitz, J. (2009), *Analysis, synthesis and design of chemical processes*, 3rd ed., Pearson Education.

U.S. Office of Energy Efficiency and Renewable Energy. (2020), 'A history of geothermal energy in America'. U.S. Department of Energy, Geothermal Technologies Program. www.energy.gov/eere/geothermal/history-geothermal-energy-america.

Uusitalo, A., Turunen-Saaresti, T., Honkatukia, J., Colonna, P. & Larjola, J. (2013), 'Siloxanes as working fluids for mini-ORC systems based on high-speed turbogenerator technology', *Journal of Engineering for Gas Turbines and Power* **135**(4), 042305.

van de Bor, D. M., Infante Ferreira, C. A. & Kiss, A. A. (2015), 'Low grade waste heat recovery using heat pumps and power cycles', *Energy* **89**, 864–873.

van Kleef, L. M. T., Oyewunmi, O. A., Harraz, A. A., Haslam, A. J. & Markides, C. N. (2018), 'Case studies in computer-aided molecular design (CAMD) of low- and medium-grade waste-heat recovery ORC systems', in *31st International Conference on Efficiency, Cost, Optimization, Simulation and Environmental Impact of Energy Systems*, 17–22 June, Guimarães.

Velasco, S., Roco, J. M., Medina, A. & Hernández, A. C. (1997), 'New performance bounds for a finite-time Carnot refrigerator', *Physical Review Letters* **78**(17), 3241–3244.

Vélez, F., Segovia, J. J., Martín, M. C. et al. (2012), 'A technical, economical and market review of organic Rankine cycles for the conversion of low-grade heat for power generation', *Renewable and Sustainable Energy Reviews* **16**(6), 4175–4189.

Velraj, R. (2016), 'Sensible heat storage for solar heating and cooling systems', in R. Z. Wang & T. S. Ge, eds, *Advances in solar heating and cooling* Elsevier, pp. 399–428.

Venkitaraj, K. & Suresh, S. (2017), 'Experimental study on thermal and chemical stability of pentaerythritol blended with low melting alloy as possible PCM for latent heat storage', *Experimental Thermal and Fluid Science* **88**, 73–87.

Verein Deutscher Ingenieure. (2010), *VDI heat atlas*, 2nd ed., Spinger-Verlag.

Walker, G. & Senft, J. (1985), *Free piston Stirling engines*, 1st ed., Spinger-Verlag.

Wang, K., Sanders, S. R., Dubey, S., Choo, F. H. & Duan, F. (2016), 'Stirling cycle engines for recovering low and moderate temperature heat: a review', *Renewable and Sustainable Energy Reviews* **62**, 89–108.

Wang, T., Zhang, Y., Peng, Z. & Shu, G. (2011), 'A review of researches on thermal exhaust heat recovery with Rankine cycle', *Renewable and Sustainable Energy Reviews* **15**(6), 2862–2871.

Watson, N. & Janota, M. S. (1982), *Turbocharging the internal combustion engine*, Palgrave.

Whitaker, S. (1972), 'Forced convection heat transfer correlations for flow in pipes, past flat plates, single cylinders, single spheres, and for flow in packed beds and tube bundles', *American Institute of Chemical Engineers Journal* **18**(2), 361–371.

White, A. J. (2011), 'Loss analysis of thermal reservoirs for electrical energy storage schemes', *Applied Energy* **88**(11), 4150–4159.

White, M., Oyewunmi, O., Haslam, A. & Markides, C. (2017), 'Industrial waste-heat recovery through integrated computer-aided working-fluid and ORC system optimisation using SAFT-Γ Mie', *Energy Conversion and Management* **150**, 851–869.

White, M. & Sayma, A. (2018), 'A generalised assessment of working fluids and radial turbines for non-recuperated subcritical organic Rankine cycles', *Energies* **11**(4), 800.

White, M. T., Oyewunmi, O. A., Chatzopoulou, M. A. et al. (2018), 'Computer-aided working-fluid design, thermodynamic optimisation and thermoeconomic assessment of ORC systems for waste-heat recovery', *Energy* **161**, 1181–1198.

Yan, Z. & Chen, J. (1989), 'An optimal endoreversible three-heat-source refrigerator', *Journal of Applied Physics* **65**(1), 1–4.

Zalba, B., Marın, J. M., Cabeza, L. F. & Mehling, H. (2003), 'Review on thermal energy storage with phase change: materials, heat transfer analysis and applications', *Applied Thermal Engineering* **23**(3), 251–283.

Zhou, D., Zhao, C.-Y. & Tian, Y. (2012), 'Review on thermal energy storage with phase change materials (PCMs) in building applications', *Applied Energy* **92**, 593–605.

Zubair, S. M. & Chaudhry, M. A. (1994), 'Exact solutions of solid-liquid phase-change heat transfer when subjected to convective boundary conditions', *Heat and Mass Transfer* **30**(2), 77–81.

Index